RARE EARTH ELEMENT

Edited by **Jose Edgar Alfonso Orjuela**

Rare Earth Element

http://dx.doi.org/10.5772/65576
Edited by Jose Edgar Alfonso Orjuela

Contributors

Lynne Macaskie, Angela Murray, Sayo Moriyama, Iryna Mikheenko, Sarah Singh, Vernon Somerset, Namil Um, Zhi-Cheng Tan, Quan Shi, Irzaman Husein, Hendradi Hardienata, Akhiruddin Maddu, Aminullah Aminullah, Husin Alatas, Andrey Svinarenko, Alexander V. Glushkov, Olga Khetselius, Valentin Ternovsky, Yuliya Dubrovskaya, Anna Kuznetsova, Vasily Buyadzhi

Published by InTech
Janeza Trdine 9, 51000 Rijeka, Croatia

Publishing Process Manager Edi Lipovic
Technical Editor SPi Global
Cover InTech Design team

First published July, 2017
Printed in Croatia
Legal deposit, Croatia: National and University Library in Zagreb

Additional hard copies can be obtained from orders@intechopen.com

Rare Earth Element, Edited by Jose Edgar Alfonso Orjuela
p. cm.
Print ISBN 978-953-51-3401-5
Online ISBN 978-953-51-3402-2

For EU product safety concerns:
IN TECH d.o.o., Prolaz Marije Krucifikse Kozulić 3, 51000 Rijeka, Croatia,
info@intechopen.com or visit our website at intechopen.com.

Meet the editor

Jose Edgar Alfonso Orjuela earned his bachelor's degree in Physics in 1987 and Magister degree in Physics in 1993; both titles were from Universidad Nacional de Colombia, and his PhD degree in Physics in 1997 was from Universidad Autónoma de Madrid (Spain). He has been working in materials science, especially in oxides of molybdenum and lithium niobate doped with rare earths. In these materials, he has studied the thermal coefficient expansion, luminescence emission, and magnetic behavior. His other works have focused on electrical and optical properties of materials grown in thin films. Currently, he is growing multifunctional films in order to characterize their optical and electrical properties.

Contents

Preface

The rare earths (REs), also called lanthanides, which means "To lie hidden," include elements from lanthanum (La) (Z = 57) through lutetium (Z = 71). These elements have physical and chemical properties that allow them to be used in different scientific and technical applications.

Some physical characteristics of the REs have been associated with their atomic electric configuration that allows hybridization of the 6s, 5d, and 4f orbitals. For instance, the hybridization of these orbitals is higher in cerium, which has the lowest melting point (1065 K) of all RE elements. Additionally, hybridization determines the metallic radii that influence density and hardness. In the case of the REs europium stands out because it has the lowest density in the series at 5.24 g/cm^3 and the largest metallic radius in the series at 208.4 pm.

Moreover, the resistivity of the REs has a value of 29–134 μΩ-cm, which is higher than in metallic elements such as aluminum, which have resistivity values of 2.655 μΩ-cm. The magnetic properties of the REs can be explained by the presence of electrons in the f orbital not unpaired.

The chemical properties of lanthanides are explained by 4f orbitals, which are usually filled. These orbitals are depicted as: f_{z^3}, f_{xz^2}, f_{yz^2}, f_{xyz}, $f_{z(x^2-y^2)}$, $f_{x(x^2-3y^2)}$, and $f_{y(3x^2-y^2)}$, which in quantum mechanics is called the "cubic set" or general set. The 4f orbitals penetrate the core and are isolated, and thus they do not participate in chemical bonding. This explains why crystal field effects are small and why they do not form π bonds

Because of these physical and chemical characteristics of the rare earth elements (REEs), many technologies develop devices with low weight, reduced emissions, and minor energy consumption, and obtain greater efficiency, performance, miniaturization, speed, durability, and thermal stability. All these advantages have aroused scientific interest in themes that comprise exploitation and the recovery of REEs. Therefore, the two first chapters of this book are dedicated to recovery of the REs by means of two methods. The first one is acid leaching as a hydrometallurgical method to obtain the REEs from waste containing them; the effective application of the acidleaching process is discussed through the leaching behavior of REs in acidic solution and the synthesis method—REs during acid leaching for recovery from waste. The second method to recover REs from wastes and ores uses two approaches that involve biogenic phosphate. The first uses an enzyme that biomineralizes RE phosphates into crystals in the extracellular polymeric matrix (EPM). The enzyme is also localized in the EPM to provide a continuous phosphate feed, which results in extensive biomineralization. The bacteria can be immobilized in a column, which allows continuous metal removal, even at high flow rates. The second approach for RE recovery uses hydroxyapatite to capture RE^{3+}, which is localized at the grain boundaries of the small biocrystallites and then substituted by Ca^{2+}. Thereafter, RE is recovered magnetically.

The next two chapters deal with spectroscopy of the REEs, specifically Chapter 3, which presents and reviews the results of studying spectra, autoionization resonance characteristics of ytterbium and thulium. Computing spectra, autoionization resonance parameters is carried out within the relativistic many-body perturbation theory (PT) and generalized relativistic energy approach. Additionally, the chapter presents a brief review of the theoretical and experimental works on spectroscopy of certain lanthanide atoms. These results are of great interest to the development of atomic spectroscopy with different applications, such as in plasma chemistry, astrophysics, laser physics, and quantum electronics, and Chapter 4 shows the results obtained in determining the concentration of cerium, La, and praseodymium employing inductively coupled plasma optical emission spectroscopy. The optical spectroscopy method is selective, sensitive, and applicable for REE determination in environmental samples with complicated matrix effects.

Chapter 5 describes the sintering of the following RE complex: $Eu(Glu)(Im)_5(ClO4)_3{\cdot}3HClO_4{\cdot}6H_2O$,$Nd(Gly)_2Cl_3{\cdot}3H_2O$ and $Eu(Glu)(Im)_5(ClO_4)_3{\cdot}3HClO_4{\cdot}6H_2O$ from amino acids. The thermodynamic property studies on these complexes were performed via low-temperature heat capacity, standard molar enthalpy of formation was determined by isoperibol solution reaction calorimetry phase transition, and thermodynamic functions were determined by adiabatic calorimetry. For the third complex, microcalorimetry was used to investigate the interaction between the complex and *Escherichia coli*.

Finally, Chapter 6 explains that La when incorporated as dopant in the crystallographic lattice of ferroelectric films changes their structural, electrical, and optical properties significantly. The changes can be measured since, for instance, barium strontium titanate (BST) films doped with La can change the material electric properties from insulator to semiconductor.

Dr. Jose Edgar Alfonso Orjuela,
Materials Science and Surfaces Group,
Physics Department,
Universidad Nacional de Colombia,
Colombia

Section 1

Rare Earths Recovery

Chapter 1

Biotechnology Processes for Scalable, Selective Rare Earth Element Recovery

Lynne E. Macaskie, Sayo Moriyama,
Iryna Mikheenko, Sarah Singh and Angela J. Murray

Additional information is available at the end of the chapter

http://dx.doi.org/10.5772/intechopen.68429

Abstract

Biorecovery of rare earth elements (REE) from wastes and ores is achieved by bacteria using biogenic phosphates. One approach uses an enzyme that biomineralises REE phosphate crystals into the extracellular polymeric matrix (EPM). The enzyme, co-localised in the EPM, provides a continuous phosphate feed into biomineralisation. The bacteria can be immobilised in a column, allowing continuous metal removal. Metals biocrystallise at different rates. By choosing suitable conditions and column flow rates selective recovery of REE against uranium and thorium can potentially overcome a bottleneck in recovery of REE from mine tailings and ore leachates co-contaminated with these radionuclides. Another approach to REE recovery first lays down calcium phosphate as hydroxyapatite (Bio-HA) using the enzymatic process. Bio-HA then captures REE, loading REE of up to 84% of the HA-mass. REE^{3+} first localises at the grain boundaries of the small bio-crystallites and then substitutes for Ca^{2+} stoichiometrically within the HA. After REE capture the bio-HA/REE hybrid can be separated magnetically. A wider concept: using a 'priming' deposit of one mineral to facilitate the capture of REEs, has been shown, potentially providing a basis for selective REE recovery which would provide advantages over the > 100 steps currently used in commercial REE refining.

Keywords: bacteria, hydroxyapatite, immobilised biofilm, rare earth recovery, rare earth selectivity

1. Introduction

Rare earth elements (REE), listed as 'critical materials' by the EU, are essential for modern electronics and for meeting global commitments to 'green' technologies, e.g. renewable energy (i.e. wind, solar) and energy efficient vehicles [1]. World REE production has approximately doubled in the last decade; consumption was forecast to reach 210,000 T in 2016 but >95% of global REE supply is from China, causing major geopolitical (e.g. soaring prices, restricted export quotas) and environmental issues. Demand for some REE will outstrip supply, leading to economic concerns about security of supply and an urgent requirement to find alternative sources. REE-bearing environments are ubiquitous but are considered too low-grade to be economically extracted or they may contain toxic and/or radioactive elements, complicating REE processing; more than 100 steps are used in current REE purifications.

Currently, hydrometallurgy processes are used to dissolve, separate and purify REE from ores [1]. A widely used method, saponification, produces more than 20 million tonnes of wastewater annually; the scale of the environmental contamination in China has led to substantial remediation bills ($0.25 billion per year) and a warning from the Chinese Ministry of Environmental Protection for REE producers to clean up emissions or face being closed down [2]. Ore extraction has moved towards bio-mining (i.e. bio-leaching/bio-beneficiation), a potentially more economical, environmental friendly approach [3], but it is often overlooked by industry due to longer extraction times, difficulty in controlling microbial reactions and unsuitable ores/materials. Microorganisms are well known to take up REEs [4–9] and can also control the environmental form, mobility and transport of metals [10]. Hence, harnessing biotechnological methods for REE extraction/recovery could lead to more economically viable, environmental friendly processes. For example while inorganic acids can selectively leach REE from clay materials [11], organic acids (such as produced by microbes) were shown to selectively extract REE from phosphate-minerals (apatite/monazite) [12] and soils [13]. Ligands with carboxyl functional groups (similar to the active groups of organic acids as found on microbial cell surfaces and in extracellular polymeric materials) have been shown to be efficient, reusable, selective leaching/extracting agents [14], which could overcome issues associated with the use of toxic additives and the generation of large waste streams [1]. Reactive minerals (often made biogenically) have remarkable morphologies and/or improved functionality for metal removal [15], for example nano-structured MnOx has 400 times higher uptake than commercial material for the removal of contaminants [16, 17], Hydroxyapatite (HA) made in the presence of organic materials can take up Cu(II) to a ten-fold higher extent than its 'chemical' counterpart [18] and biogenic HA can take up Sr^{2+} from saline solution where its chemical counterpart cannot [19]. Alternatively, bacteria can simply synthesise REE phosphate enzymatically to metal loadings of several-fold higher mass than the biomass itself [20, 21]. The neo-mineral is nanoform (14 nm crystallites) [21, 22] which has implications for onward facile processing; with a much larger surface area than large mineral crystals it would be easier to generate a leachate under benign conditions for easier metal recovery from a resulting concentrate.

The purpose of this review is to highlight developments towards a biotechnological approach to REE recovery from solution and to indicate where future opportunities (and difficulties) may lie in a bio-refinery concept towards cleaner, more efficient REE recovery and processing.

The examples described centre upon the use of biogenic REE phosphates, some of which have particular applications, e.g. as catalysts in the manufacture of butadiene rubber (Nd) [23] and as optically active materials (Eu) [24, 25]. The concept of 'one-pot bio-refining' of new material from waste brings new opportunities for application of combinatorial chemistry and synthetic biology which cannot be achieved by chemistry alone and key bottlenecks to progress in novel REE recovery and processing are highlighted.

2. Biotechnology for recovery of rare earth elements from solution

It is well known that microbial cells have a strong affinity for REEs [4–9]. As well as by simple biosorption bacterial systems can accumulate metals as heavy bio-mineral deposits, e.g. see Macaskie et al. [26]. Metallic ions (e.g. UO_2^{2+}, La^{3+}) can be converted into metal phosphate minerals (HUO_2PO_4, $LaPO_4$) enzymatically. Using a well-documented paradigm system [26] a *Serratia* sp. (strain N14; formerly designated as a *Citrobacter* sp.) can, especially under carbon-restriction, up-regulate an acid phosphatase. The corresponding *phoN* gene, under control of the *phoP/phoQ* regulon, is up-regulated in response to various cell stresses. The phosphatase is 'trafficked' to and through the cell surface layers (periplasmic space and outer membrane) and locates in the extracellular polymeric matrix (EPM) [27] which comprises heavily hydrated polysaccharide chains and some lipids derived from cell membrane material [27]. The EPM tethers and supports the phosphatase (**Figure 1A**) and, when supplied with a suitable organic phosphate molecule, the enzyme liberates inorganic phosphate to precipitate with metal ions. This forms nucleation sites. Hence, the EPM scaffolds the growing metal phosphate bio-mineral (**Figure 1B**) that grows with addition of further metal and phosphatase substrate. In the case of a biofilm (**Figure 1C**), heavy mineral deposits eventually coat the cells and hence the biofilm (**Figure 1D**). Importantly, the bacteria do not need to be alive as the enzyme is essentially immobilised; the need for metabolism ceases when the enzyme is made and exported into the EPM. The phosphatase activity is stable and hence heavy mineral deposits are obtained (**Figure 1D**). Bacteria readily grow as biofilms on surfaces, such as reticulated polyurethane foam (**Figure 1E** and **F**) 'tethering' the bio-mineral (**Figure 1G**) against the flow within flow-through columns (**Figure 2**). This approach to metal recovery was initially developed for removal of radionuclides from solution where lanthanides such as La^{3+} and Eu^{3+} form a 'surrogate' for trivalent transuranic elements [20, 28, 29]. Effective, continuous metal removal was reported using La^{3+} using cells entrapped in a gel [20]. La is not a highly valuable REE and hence development of REE bio-recovery technology has more recently focused on the removal of Nd^{3+} and Eu^{3+}.

Early work used a gel-immobilisation system with bacteria grown in batch culture as free cells and then immobilised [20]. Shredded gel was placed into a glass column for use as a continuous filter. Later studies using biofilm-immobilised cells (**Figures 1G** and **2**) used columns comprising eight biofilm-coated polyurethane reticulated foam discs stacked in a glass column [21] with solution pumped upwards (flow rates as required). The solution comprised rare earth ions (REE^{3+}), glycerol 2-phosphate (5 mM; phosphate donor to yield inorganic phosphate for the growing mineral) and citrate buffer (2 mM) to the required pH. Citrate provides a known amount of metal chelation. This is important to compare the recovery of different metals in an identical solution

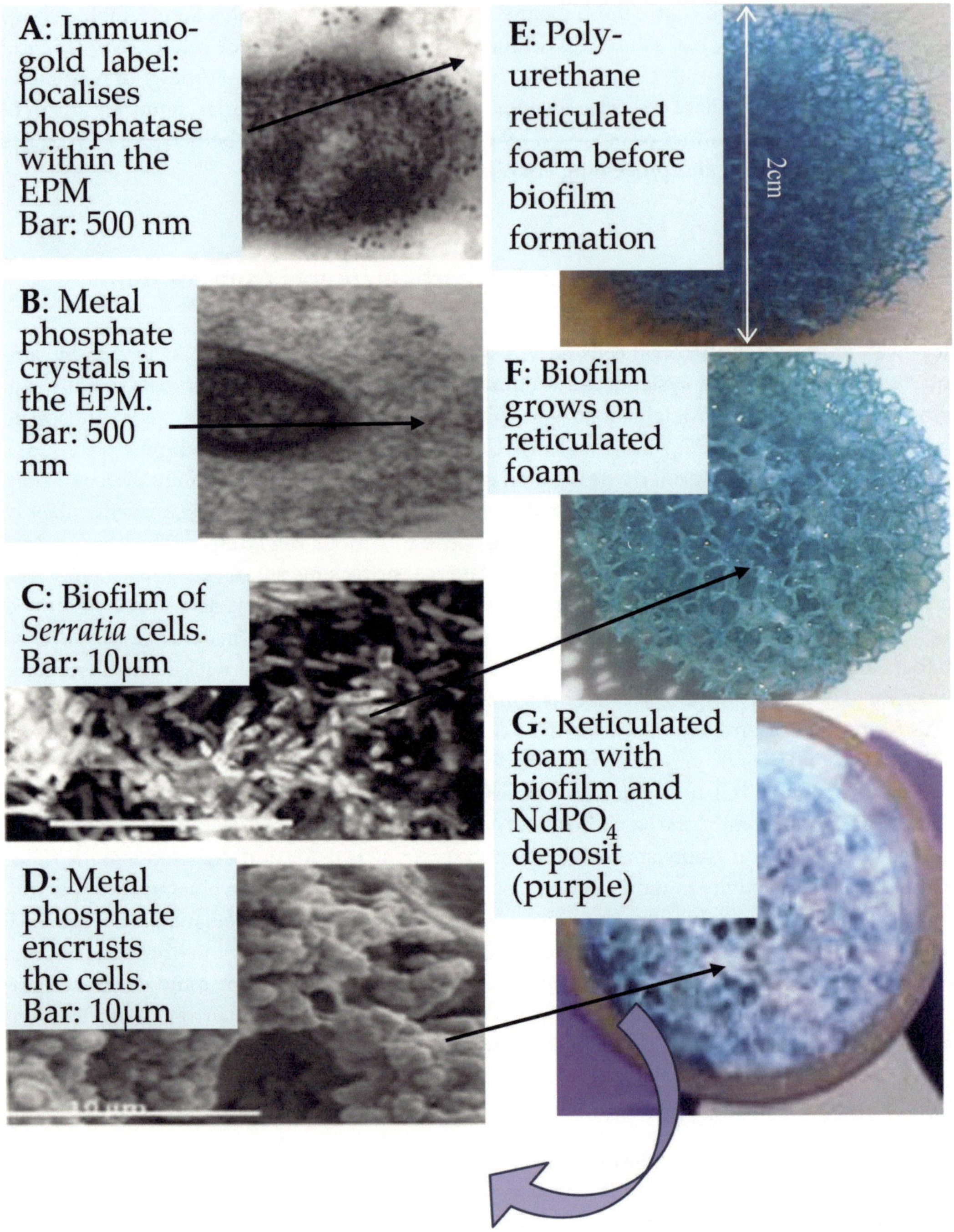

Figure 1. Phosphatase location within the extracellular polymeric material (A) and metal phosphate deposition in EPM (B) by *Serratia* sp. cells. (C and F): Biofilm on reticulated foam discs. (E): foam disc before biofilm colonisation. (D and G): Deposition of $NdPO_4$ after flowing a solution of 1 mM Nd^{3+} and 5 mM glycerol 2- phosphate through a column (seen end-on in G). Blue arrow: Disc as shown in Figure 2.

http://dx.doi.org/10.5772/intechopen.68429

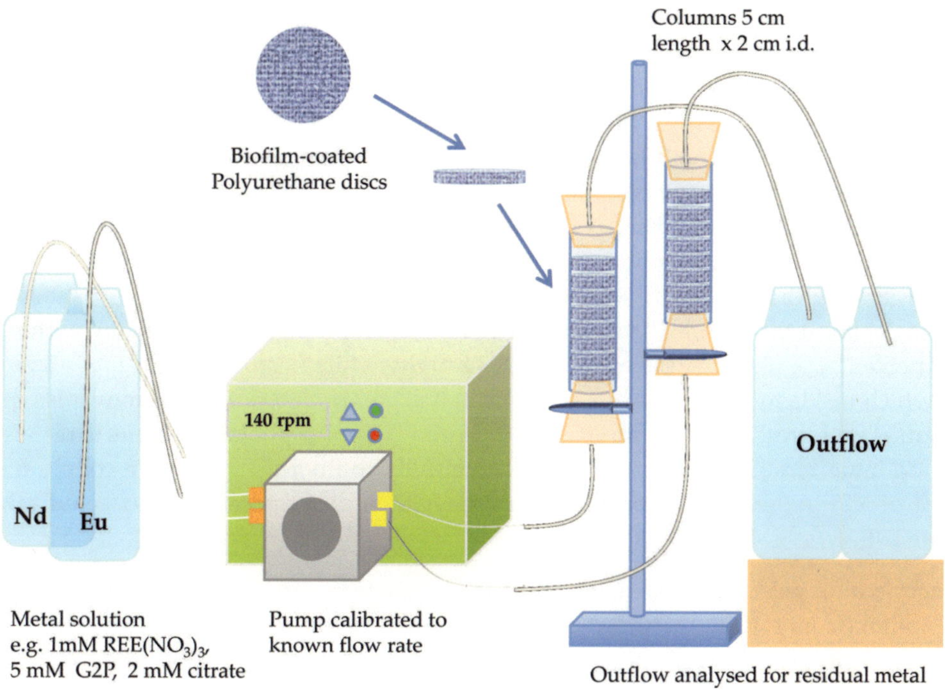

Figure 2. Experiment setup for REE recovery. Column activity is expressed as the flow rate-activity relationship, i.e. that flow rate giving 50% removal of metal under the set conditions. Flow rate-activity ($FA_{1/2}$) value for Nd^{3+} and Eu^{3+} removal by columns was 273 mL/h [21] (see text).

matrix. For example, in the case of U(VI) (and especially Th(IV)) metal hydrolysis occurs extensively with non-chelated metal. This (i.e. loss of OH^- from the solution) promotes a fall in the solution pH as well as formation of metal hydroxide colloids; to forestall hydrolysis and ensure that the experiments were comparable the citrate (as citrate buffer) maintained the required pH.

A working column can be defined in terms of its flow rate activity relationship. A very useful parameter is the $FA_{1/2}$ value which is defined as the flow rate that maintains 50% of the metal removed from the solution. The activity of a column is accurately described as a plug flow reactor and it can be quantified via an integrated form of the Michaelis-Menten equation. This describes the immobilised enzyme component with, in addition, an 'inefficiency' factor which is a lumped parameter describing the additional combined effects of the dissociation of the metal from its citrate complex and metal removal from solution into the precipitate; the latter can be viewed as the propensity of the metal to form a phosphate deposit [30]. However, the actual times taken to attain the multiple equilibria need not be known; they are dynamic in the flow-through system. Some relevant data are reported in non-flow-through systems [31]. The result observed for column activity is the net outcome as a measure of metal removed under a particular condition. In some cases competing reactions may occur. For example, the phosphatase showed transphosphorylation activity (i.e. to regenerate substrate) as shown by using ^{31}P nuclear magnetic resonance (NMR) spectroscopy [32]. It was found by trial and error that columns optimally require a substrate to metal ratio of more than 2:1. The predictive model was proved in an example of uranium removal, firstly from test solution [30] and then from acidic wastewater from a uranium mine which also contained excess sulphate ion [33]. Importantly, the degree of interference by other components of the flow, such as excess sulphate (or nitrate), can be overcome simply by slowing the flow rate by an amount which can be predicted in a given system [34, 35] to maintain the metal removal at a high level. The effect of the solution pH can be established in a similar way [36]. From this modelling the size of the column required for a particular application can be projected assuming that the major solution components and the pH are known. It should also be borne in mind that, while the activity of different cell preparations differs by within only a few percent, the phosphatase enzyme loses activity while in storage (half-life is 3 months [21]) which needs to be factored-in at the design stage. In practice, a column series would be optimal with fresh columns inserted as required. The phosphatase enzyme is resistant to common inhibitors [37] but has a marked sensitivity to fluoride which is a common component of REE minerals (e.g. bastnäsite, synchysite) and the recently discovered mineral parisite [38].

Polyacrylamide gel-immobilised free cells (pre-grown in carbon-sufficient batch culture) and biofilm (from carbon-restricted culture) were compared with respect to the removal of REE^{3+} from solution. The phosphatase activity is 10-fold less for free cells grown batch-wise than for biofilm cells (250 and 2500 nmol product/min/mg protein). The $FA_{1/2}$ value for REE^{3+} removal was 3.9 and 45.6 mL/h/mg cells, respectively, i.e. an underperformance of ~15% by the gel-immobilised cells which suggests some diffusional constraint within the gel.

3. The need for selective REE recovery against Th(IV) and U(VI) and potential development of the bio-recovery method to metal mixtures

A barrier to exploitation of historic uranium mine tailings (and indeed bottlenecks in primary refining of REE from many minerals) is their widespread occurrence with U and/or Th. For example, at the Elliott Lake site in Ontario, Canada, emphasis has diversified from primary U production to embrace also the untapped REE resource. In 2014, Pele Mountain Resources expanded its business model to include the sustainable development of a low-cost, early-to-market, rare earth processing centre at Elliot Lake (designated as the Eco Ridge Mine Rare Earth and Uranium Project: 'Eco Ridge'); however, this is capital-cost intensive while the price of rare earths has fallen. Hence, an initial monazite processing strategy was adopted (since 2015) for project development at Eco Ridge as the proposed site for Canada's first centre for processing both rare earths and uranium as a low cost alternative to mine development, pending development of sustainable rare earth production and supply chains [39]. Recent market conditions (e.g. a decline in rare earth prices) do not support the development of new hard-rock mining projects *per se* but REE separation is required for the future recovery of REE materials from wastes; hence, establishing rare earth production and separation in countries with primary resources of REE is a first step to supply chain setup.

The Eco Ridge site will start rare earth production by processing imported monazite (REE phosphates), where the high rare earth grades would justify REE production from relatively low ore tonnages. Such an initial 'processing-only' setup would reduce capital outlay and time-to-market as compared to the development of new mine processing operations since the monazite processing facility would then underpin scalable REE production from local resources. Metallurgical techniques for processing monazite are well established (although monazites are fairly resistant to dissolution), and an economic model envisages moving on from monazite to mine tailings.

The foregoing illustrates that 'cleaner' methods to separate REE and U (e.g. in mine wastes) and REE and Th (e.g. from monazite) are needed since current processing methods require multiple hydrometallurgical steps with high consumption of acids and alkalis. It should also be noted that while 'geological' monazite is highly crystalline and hence requires strong acid for dissolution, biogenically recovered monazite comprises nano-crystals of ~14 nm [21] offering a large surface area as compared to native mineral crystals. To re-manufacture monazite biogenically from solution may appear counter-intuitive but the biogenic material would require less onward processing than the primary resource and biological intervention could also offer the potential to separate REE from U and Th.

Initial tests [40] used polyacrylamide gel-immobilised cells, taking into account that both U and Th form tight complexes with citrate (which is needed to prevent the formation of colloidal metal hydroxide deposits). Natural ligands in real systems that fulfil this function would include various organic ligands arising from microbial activities in addition to geopolymers like humic and fulvic acids. Mixed Th(IV)/La(III) and U(VI)/La(III) solutions were passed through columns (25 mL volume) by the method of Tolley et al. [20] at rapid flow rates. Assuming that the time required for the metal-citrate dissociation equilibrium to re-establish following metal removal from the flow into the mineral phase is longer than the flow residence time, it was predicted (according to the relative solubility products, for example this being ~7 orders of magnitude less for the uranium mineral autonite as compared to monazite [9]) that La(III) would be removed effectively while Th(IV) and U(VI) may not. For example, at a flow rate of 198 mL/h (LnF = 5.28 mL/h), the cleavage of the glycerol 2-phosphate substrate was 33%, to give an available phosphate concentration of 1.7 mM. In the metal mixtures, the total metal was 1 mM (0.5 mM of each metal; the metals are described in molar terms to account for the differences in their mass and provide equal concentrations). **Figure 3A** shows that while 100% of La^{3+} was removed from the mixture, the removal of Th was ~20%. In contrast, in another series of tests a higher flow rate was required to 'prevent' U removal (**Figure 3B**), with only a narrow 'window' of flow rates under which 100% of La^{3+} was removed but little uranyl ion.

This narrow flow rate 'window' must be established each time since each waste would contain different components in different proportions and wastes also vary with respect to pH. Hence, while rejection of Th(IV) by the column, and hence metal separation, is feasible using column retention time alone (**Figure 3A**), a better selectivity of REE(III) against U(VI) is required in

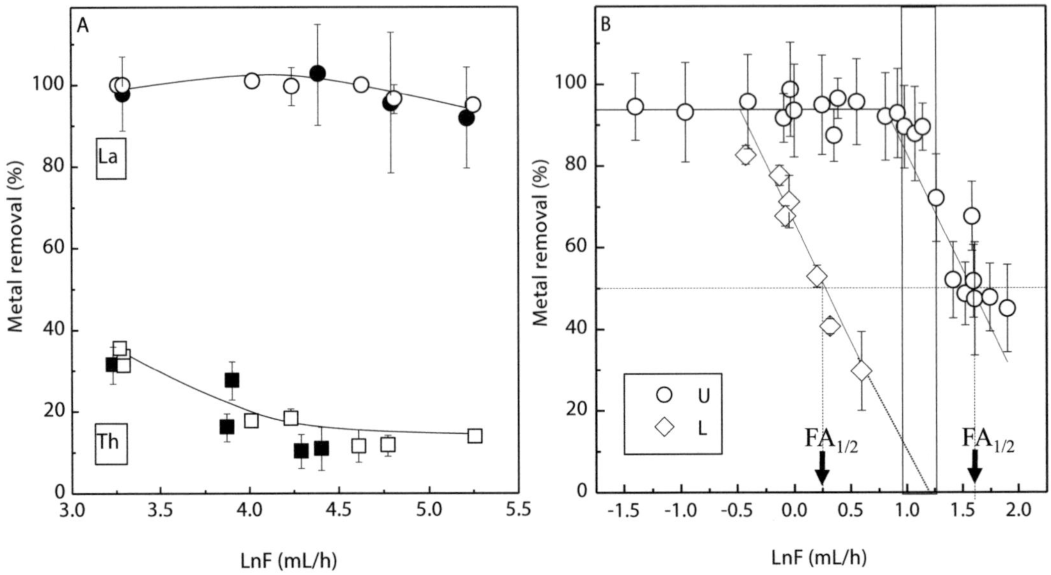

Figure 3. Flow rate-activity relationships for removal of (A) a mixture (0.5 mM of each) of La(III)/Th(IV) and (B) 1mM, separately, La(III)/U(VI) using polyacrylamide gel-immobilised cells in two series of experiments [35]. Closed and filled symbols represent metal only (1 mM) or metal in the mixture (0.5 mM of each). Good separation of La (●, ○) and Th (■, □) is shown in (A). (B) shows a narrow flow rate 'window' at which La^{3+} removal (○) was maintained while UO_2^{2+} removal (◇) was less than 10% (boxed). The concentration of glycerol 2-phosphate was 5 mM (5:1 excess over metal). $FA_{1/2}$ is flow rate giving 50% removal of metal from the flow (arrowed).

order to expand the 'window' of flow rates (**Figure 3B**) between which the former but not the latter is removed. In this respect, an additional complexing agent for UO_2^{2+} could retard its capture within the flow residence time in the column, thereby shifting the flow rate-activity relationship for U(VI) to the left, while leaving the REE^{3+} removal profile unchanged.

4. Biotechnology progress towards selective recovery of REE: the use of biofilm-immobilised cells

The early work used bacteria immobilised in polyacrylamide gel (PAG) [20, 40]. However, gel-systems have insufficient mechanical strength for large scale processes. The *Serratia* cells readily form a biofilm on solid matrices (above) which was used in the removal of $^{235/238}UO_2^{2+}$ from a real nuclear waste [41] as well as Nd^{3+} (**Figure 1G**). Hence, the possible application to selective metal recovery was examined.

Initial comparisons were made using 1 and 5 mM glycerol 2-phosphate (G2P). **Figure 4** shows that, in contrast to gel-immobilised cells, a biofilm-column gave less metal separation e.g. at a removal of UO_2^{2+} of 9.4% (at LnF = 4.96 mL/h) the corresponding removal of Nd^{3+} was 54.2% (at LnF = 4.95 mL/h). This could be attributed to the removal of a selective permeability barrier

http://dx.doi.org/10.5772/intechopen.68429

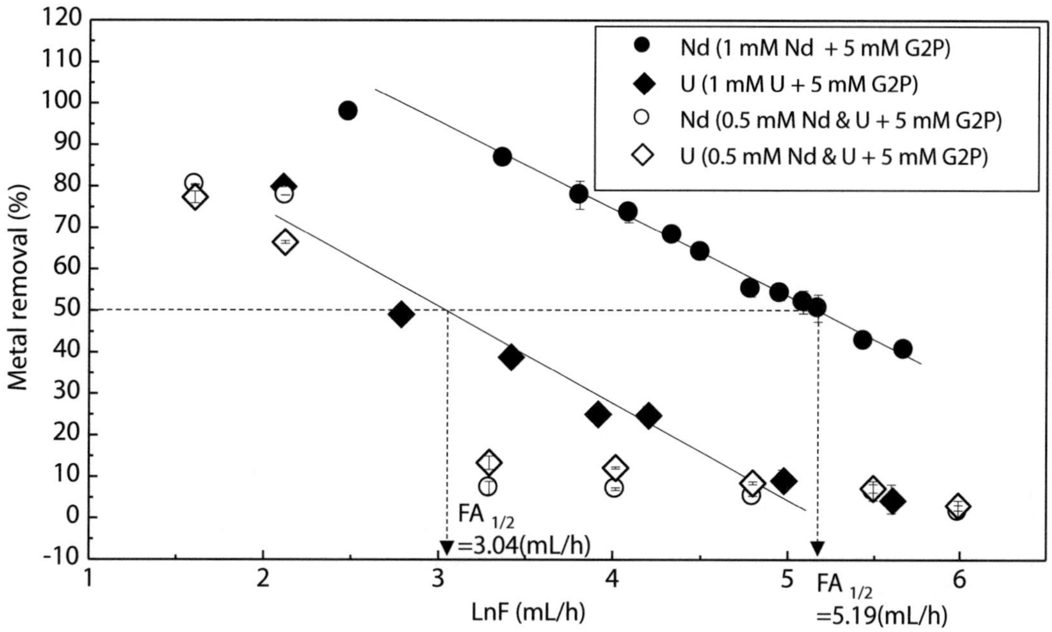

Figure 4. Removal of 1 mM Nd^{3+} (●) and 1 mM (♦) UO_2^{2+} individually from the flow (2 mM citrate buffer pH 5.5; average of two independent biofilm preparations) in separate columns with 5 mM glycerol 2-phosphate (filled symbols and from a mixture containing 0.5 mM of each metal (open symbols). The boxed area is the region of flow rate that confers some selectivity (c.f. Figure 3) as indicated by single metal tests.

in the gel that retarded UO_2^{2+} more than REE^{3+}, to a better permeability of uranyl ion through the natural permeability barrier of the extracellular polymeric materials that characterise a biofilm, or to a combination of both factors.

Removal of Nd^{3+} and UO_2^{2+} in competition with each other is also shown in **Figure 4**. The presence of excess liberated inorganic phosphate was comparable in all cases (not shown) but the mixed metal test shows no selectivity. This was confirmed using 1 mM G2P, i.e. making less free phosphate available (in order to 'starve' UO_2^{2+} of the phosphate it needs to precipitate effectively) did not selectively retard UO_2^{2+} removal.

Comparing **Figure 4** with **Figure 3B**, this study with single metals indicates that biofilm-cells are less effective than gel-immobilised cells in terms of the potential for metal separation. At the $FA_{1/2}$ value for Nd^{3+} (LnF = 5.15 mL/h) the removal of UO_2^{2+} was >20% (**Figure 4**) whereas using gel-immobilised cells the corresponding removal of La^{3+} was ~100% (**Figure 3B**). However, inspection of unpublished data of Tolley [40] (c.f. **Figure 4**) revealed that the removal of La^{3+} by PAG-immobilised cells was depressed to the same level as UO_2^{2+} in the mixture although La^{3+} removal from a single metal solution was >90% at the same flow rate. Hence, clearly the depression of REE removal in the presence of UO_2^{2+} (**Figure 4**) is not attributable to the choice of cell immobilisation method (and hence not attributable to the extracellular polymer produced by biofilm cells). These results show similarities with earlier data of Ohnuki et al. [42].

Here, while Ce(III) was removed onto cells of *Pseudomonas fluorescens* in the presence of citrate comparably to Pr(III) at acidic pH, at pH 6 and above Ce(III) removal was reduced by ~40-fold as compared to La(III) concomitantly with an oxidation of Ce(III) to Ce(IV) observed by XANES spectroscopy. A similar study [42] examining the uptake of Eu(III), Th(IV) and Pu(IV) in the presence of desferrioxamine B (DFO) as a complexing ligand showed that, at pH 6, the removal of Eu(III) was ~6-fold greater than for Th(IV) while the removal of Pu(IV) was ~100-fold less than for Th(IV). Analysis of the oxidation state of Pu was not attempted but it seems possible that partial oxidation of Pu(IV) to Pu(VI) may has occurred in an analogous way to that of Ce(III). In both the Ce(III)/(IV) and Pu(IV)/(VI) tests removal of the mixed valence species was drastically reduced as compared to a single valence species alone. However, this study [42] did not use a mixed metal challenge system, nor was any attempt made to establish the proportions of each metal species in the putative mixture for a single element, nor of the 'preference' of the cells for a particular metal species of that element.

An additional factor in the current study is that the removal of both metals from their mixture failed at rapid flow rates (**Figure 4**) although excess available Pi was confirmed (>4 mM). It is suggested that a large polymeric complex may be formed between the 2 mM citrate and 0.5 mM Nd^{3+}/0.5 mM UO_2^{2+} such that their availability into solution becomes curtailed and metal release into solution may not occur within the column residence time at rapid flow rates. Indeed, in the Pu/DFO/*Ps. fluorescens* system Pu removal was decreased by a further order of magnitude by increasing the Pu/DFO ratio from 1:1 to 1:100 [42], so that less available Pu would be available for interaction with the cells.

Multi-metal polymeric citrate complexes are now well known (e.g. [43]) and can be included in the class of metal-organic frameworks (MOFs) that are gaining increased recognition. Importantly, the structure of the MOF depends on the metals within the complex; a multi-metal citrate framework was shown to assume a cubic crystal structure [44]; many MOF materials have been described [45]. In conclusion, although immobilised bacteria removed metals individually, when metals are combined in the presence of complexing ligands the REE is 'withdrawn' from availability, being captured into a putative bimetallic MOF. Hence, another agent is required that has the potential to compete with putative MOF formation selectively for one metal over another.

5. Biotechnology progress towards selective recovery of REE: use of tributyl phosphate (TBP) as a possible selection tool

The well-established 'PUREX' process uses tributyl phosphate (TBP) to sequester actinides against a background of metal cations. An early study indicated that the complexation of TBP to uranyl ion was more than to REE [46]. Hence, TBP was considered as a possible agent to retard removal of UO_2^{2+} in the REE-accumulating column to enable later recovery of U(VI) downstream. In addition, while effective removal of UO_2^{2+} from acidic minewater at the expense of glycerol 2-phosphate was achieved successfully [33] a barrier to industrial implementation remains the cost of G2P at scale [47]. This would impact adversely upon commercial REE recovery. As a potential alternative phosphate donor, TBP is cheap and readily available. These dual advantages prompted evaluation of TBP as an alternative phosphate donor, this requiring, probably, the concerted action of phospho tri-, di- and mono-esterases. On the other hand, since the dibutyl phosphate complex of uranium is insoluble [48, 49] only one enzymatic step might be required, giving DBP (**Figure 5**).

Uranium removal (90%) from acidic U-minewater supported by TBP was achieved by a mixed culture [50]. The column input and output concentrations of TBP were 1.2 and 0.8 mM, respectively; correspondingly, 0.4 mM UO_2^{2+} was removed but the flow rate was very slow (<1 mL/h). The accumulated solid was confirmed to contain HUO_2PO_4 [50] but the presence of other solid uranium species was not sought. As a potential tool for metal separation it is important that TBP does not impact negatively upon REE removal. As confirmed by **Figure 6**, removal of Nd^{3+} was indistinguishable between columns with and without TBP, while the stability of TBP against PhoN phosphatase was shown by an identical release of phosphate under both conditions (**Figure 6**).

Controls established no phosphate leaching from the tubing or materials. With each new flow rate the column was pre-equilibrated before sampling. To check that columns did not require further time to incorporate TBP, up to 20 column volumes were tested at each flow rate. Between experiments, columns were returned to the initial condition (no TBP) to establish stability for re-use in subsequent tests. Data were normalised to factor-in the decay time of the biofilm phosphatase between tests [21]. Total Pi release (open symbols) was calculated as equivalent to Nd^{3+} removal (the solid was $NdPO_4$ [21]) plus that found by assay of soluble Pi. The solid and soluble phosphate data were pooled to give total phosphate liberated. This experimental design was adopted in subsequent tests.

Since TBP was shown to support U-removal by a mixed culture (at a flow rate of 1 mL/h as compared to >200 mL/h by using G2P, above), it may be expected that a tight binding of uranyl ion to TBP would slow the rate of the precipitation reaction with phosphate liberated

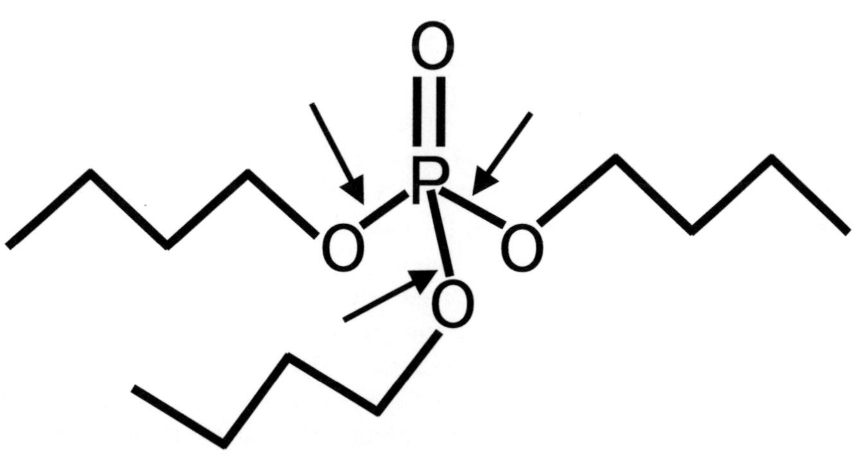

Figure 5. The structure of tributyl phosphate (TBP). Note that, unlike with glycerol 2-phosphate (G2P), three phosphoester bonds (arrowed) require cleavage to yield free inorganic phosphate (Pi) via intermediate DBP. The products would be 1 mol of Pi and 3 mol of butanol. The enzymology is unknown.

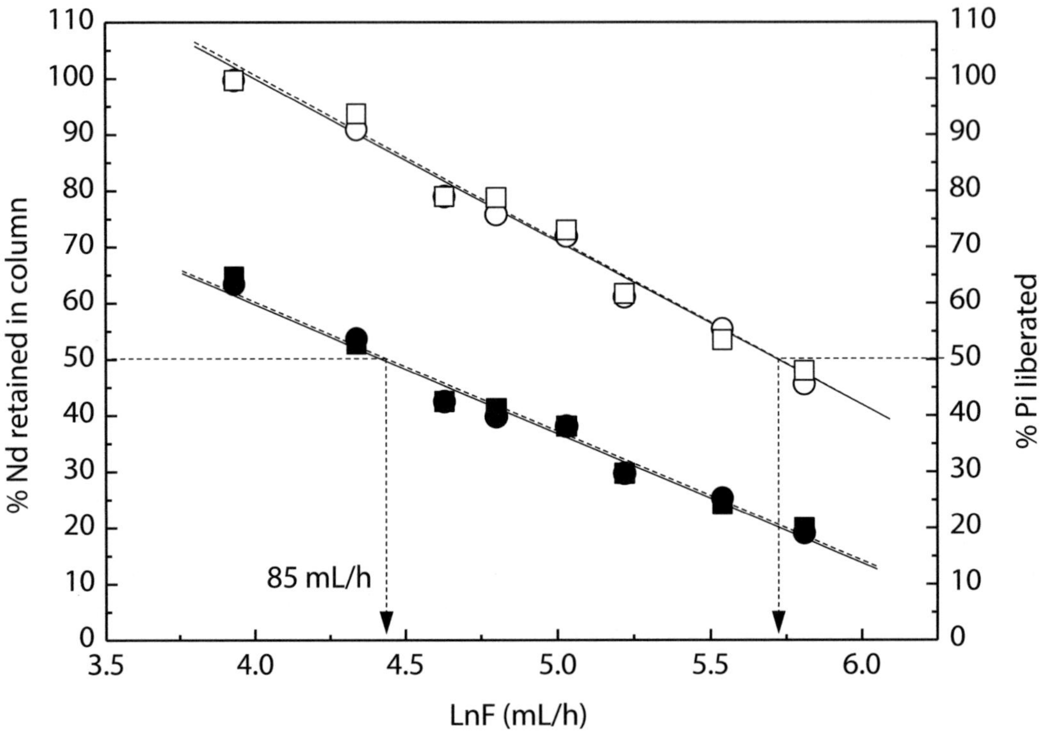

Figure 6. Effect of TBP on Nd^{3+} removal. Biofilm-foam discs (8) were packed into glass columns (Figure 2). Each column received an equivalent mix of discs from positions in the growth vessel. Tests used biofilms made from two preparations (variation <5%). Solution (1 mM G2P, 1 mM Nd^{3+}, 2 mM citrate, pH 5.5 (●, ○)) was passed upwards at varying flow rates. TBP (2 mM Alfa Aesar, fresh), was supplemented directly into the solution and agitated vigorously to maintain an aqueous suspension (■, □). Open symbols: Pi liberated. Filled symbols: metal removed.

from G2P in the bulk solution (by maintaining a low concentration of free UO_2^{2+}), but, bound to TBP, the uranyl would be held adjacent to any liberated DBP for removal in a slower flow rate column downstream.

An early study indicated the ability of cells of *Serratia* sp. strain N14 to hydrolyse TBP [51]. However, the phosphatase also behaved as a transphosphorylase regenerating alkyl phosphate substrate from the alcohol product in a futile cycle [32] in the absence of a 'phosphate trap' such as a heavy metal ion. However, the breakdown of TBP was genetically unstable [52]. Later work [53] showed that degradation of TBP by *Serratia odorifera* occurred via co-metabolism of a primary substrate that involved neither the release of detectable di-butyl or mono-butyl phosphates nor of Pi. These authors subsequently identified an oxidoreductase activity in *Rhodobacter palustris* that was implicated in TBP breakdown, also noting that oxidoreductases do not necessarily require molecular O_2 but can utilise the oxygen present in water [54]. In this context two salient features prompted this current study: (i) the breakdown of TBP in highly radioactive solution (radiolysis) proceeds via chemical formation of dibutyl phosphate from TBP [48] (presumably via free radical attack from radiolysis of water); the DBP forms insoluble metal precipitate (see above) and (ii) the *Serratia* phosphatase shows bi-functional phosphohydrolase and oxidoreductase activities and was found to generate free radicals [55], this bi-functionality resembling a reported vanadium peroxidase [56].

Application of the bacteria to support metal removal via trialkyl phosphate breakdown was studied empirically in early work using Cd^{2+} as the phosphate trap [57]. The column (*Serratia* cells immobilised in polyacrylamide gel; this strain N14 was originally designated as a *Citrobacter* sp. in early work) was challenged with 2 mM Cd^{2+} with limiting concentrations of glycerol 2-phosphate: 1 or 0.5 mM; the former supported 50% removal of 2 mM Cd^{2+} (i.e. leaving 1 mM available Cd^{2+}) while 0.5 mM G2P supported ~30% of 2 mM Cd^{2+} (**Figure 7**). With 0.5 mM G2P/5 mM G2P alkyl phosphate (1:10), the removal of Cd^{2+} fell rapidly to the 'baseline' level of 0.5 mM G2P alone. A mixture of 1:4 gave a slower loss of activity in each case (**Figure 7**) with a flow of 2–3 L sustained before the 'baseline' attributed to 0.5 mM G2P alone was reached. However, a mixture of 1 mM G2P:2 mM TBP supported steady state removal of 70, 75 and 90% of the Cd^{2+} (i.e. this accounted for more phosphate consumed than was accountable as the Pi liberated from G2P) using trimethyl, triethyl and tributyl phosphates, respectively (**Figure 7**). For 90% of 2 mM Cd^{2+} to be removed (i.e. 1.8 mM) with, maximally, 1 mM Pi able to come from the G2P (i.e. 1 mM G2P provided in the inflow) the implication is that 0.8 mM Pi had been liberated from TBP (from 2 mM TBP provided).

This early study [57] offered no explanation for the observed behaviour but the requirement for G2P to facilitate TBP breakdown is in accordance with a hypothesis of co-metabolism proposed

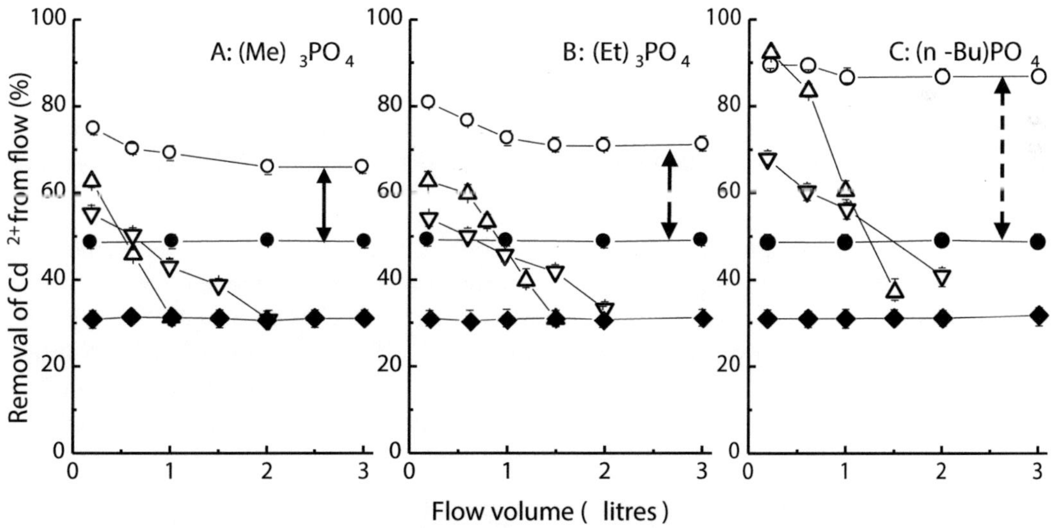

Figure 7. Ability of trialkyl phosphates to support removal of Cd^{2+} from solution by polyacrylamide gel-immobilised cells with and without added glycerol 2-phosphate (G2P). Data are redrawn from Ref. [57]. The flow rate was 30 mL/h (column working volume 60 mL) with 2 mM Cd^{2+} and 1mM (●), (50% removal of Cd^{2+}) and 0.5 mM (◆) (30% removal of 2 mM Cd^{2+}) G2P. Mixtures were also supplemented with 0.5 mM G2P/5 mM alkyl phosphate (ratio of 1:10; total phosphate was 5.5 mM) (△) or 0.5 mM G2P/2mM alkyl phosphate (ratio of 1:4; total phosphate was 2.5 mM (▽) or 1 mM G2P/2 mM alkyl phosphate (ratio 1:2; total phosphate was 3 mM) (○). Trimethyl phosphate supported removal of additional ~30% of Cd^{2+} (A: solid arrow); triethyl phosphate, of additional ~40% of Cd^{2+} (B: broken arrow) and tributyl phosphate, of additional ~80% of Cd^{2+} (C: dashed arrow) at steady-state.

by Berne et al. [53, 54]. This requires that the co-product of G2P hydrolysis, i.e. glycerol, is metabolised by the cells and this was confirmed by analysis of the exit solution of metal-accumulating columns where not only glycerol but also citrate was consumed by the cells (P. Yong, M. Paterson-Beedle and L.E. Macaskie, unpublished). Under metal-accumulating conditions, the non-growing cells accumulated large deposits of polyhydroxybutyrate as a carbon and energy stored (see later) [58]. The early work [57] showed several salient features; G2P was required for utilisation of TBP (**Figure 7**) and 1 mM TBP supported a stable column. On the other hand, 0.5 mM G2P was insufficient to sustain metal recovery (**Figure 7**). Moreover, columns were 'refreshed' by a period of TBP-free flow followed by a new TBP-supplemented cycle [57], in accordance with not only a role for co-metabolism (of glycerol) as proposed by Berne et al. [53] but to replenish PHB reserve during 'refreshment' to support the next TBP cycle. Also, addition of the proposed alcohol end-product (3-fold molar excess to alkyl phosphate) inhibited metal removal [57] in accordance with a competing transphosphorylation (above) of phosphate onto liberated alcohol [32]. The removal of Cd^{2+} supported by trimethyl and triethyl phosphate was reduced by ~25–30% by addition of methanol and ethanol, respectively, and by 20% in the case of TBP/butanol [57].

A hypothesis was formulated to widen the flow rate 'window' boxed in **Figure 3**. This assumes that REE^{3+} forms a weaker complex with TBP than does UO_2^{2+} [46]; no benefit for REE removal would be anticipated with TBP (**Figure 6**) if REE^{3+} is presumed to be not in close association with TBP or its breakdown products. Rare earth phosphate ($REEPO_4$) formation is rapid and hence occurs at high flow rates while UO_2^{2+} requires more time to form a precipitate [21]. As UO_2^{2+} complexes with TBP, it is ideally placed to capture liberated phosphate (or DBP) from TBP breakdown to form uranyl phosphate precipitate in advance of the competing transphosphorylation reactions of phosphate onto liberated alcohol. Hence, incorporation of TBP into a REE/U(VI) G2P/TBP mixture might better capture $REEPO_4$ at a rapid flow rate while rejecting uranyl, which would be available for downstream recovery as uranyl-DBP at a slower flow rate.

However, as shown in **Figure 8**, with 1 mM G2P/2 mM TBP (combination as used in the early work [57]; **Figure 7C**) there was no selectivity of metal removal and, indeed, the removal of both metals was inhibited even at slow flow rates. With 5 mM G2P/0.5 mM TBP both metals were removed at slow flow rates (below 10 mL/h) but with little or no selectivity between them. In the absence of uranyl ion, Nd^{3+} appeared to be resistant to this inhibition at all flow rates (**Figure 3**) and it may be suggested that, once they are in a presumptive bimetallic citrate-MOF, the effect of TBP applies to both metals equally.

Hence, despite promising early work using La^{3+} and UO_2^{2+} separately (above) and a successful TBP supplement being shown to boost metal recovery by gel-immobilised cells (**Figure 7**), TBP was ineffective to support metal removal from a single metal or mixed Nd/U flow using biofilm-cells for reasons still to be established. In this context, it is noteworthy that other work using triethyl phosphate as a means for generating inorganic phosphate gave dissolved phosphate at concentrations far in excess of that based on equilibrium consideration but there was no detectable mineral precipitation [59]. Both metal speciation and full mass balances would form the basis for future work to shed light on these anomalies.

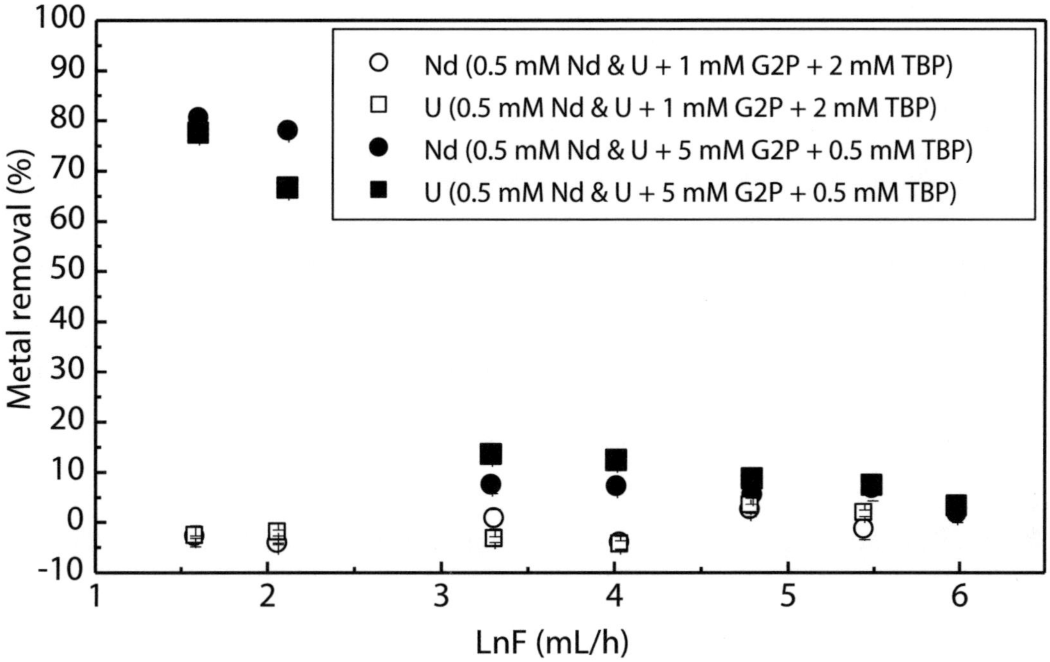

Figure 8. Removal of Nd^{3+} (●, ○) and UO_2^{2+} (■, □) from a mixture of 0.5 mM of each metal in flow with 1mM G2P/2 mM TBP (open symbols) or 5 mM G2P/0.5 mM TBP (filled symbols).

If MOF formation is confirmed, this has implications for the treatment of wastes more generally which contain organic complexing agents, e.g. those arise naturally from degradation of larger polymers [60]. In the case of uranyl removal, the use of bicarbonate ion to remove UO_2^{2+} out of the immobilised phosphate mineral in a column pre-loaded with HUO_2PO_4 was shown [30]; hence, addition of dilute bicarbonate and use of a second uranium-accumulating column could provide an alternative/supplemental route to U-retardation with downstream acidification to convert HCO_3^- to CO_2, releasing UO_2^{2+} as a free ion available for phosphate bio-mineralisation; the enzymatic bio-mineralisation is tolerant to pH down to pH 4 [33, 36] but further developmental work is required.

6. Potential for recovery of REE from bio-mineralised columns

A process for REE bio-refining into new materials would require additional processing steps. REE recovery from the bio-REE phosphate bio-material could use NH_4^+ since this has been shown for REE leaching from clay minerals and metallic scraps [61–63]. The highly nano-crystalline nature of bio-$REEPO_4$ [21] (see later discussion) would make it suitable for a more 'benign' leaching than the strong acids required for 'geological' monazite minerals. A series of tests using columns supplemented with acidic 100 mM $(NH_4)_2SO_4$ (to simulate leach liquor) showed that this had no effect on column activity at pH 5.5 (**Table 1**), with only a slight loss

	pH 5.5	pH 3.5	pH 5.5 (return)
Without/with (±) 100 mM $(NH_4)_2SO_4$	±	±	±
$FA_{1/2}$ (mL/h)	38/38	19/17	30/25

Columns were run at pH 5.5 or 3.5 with and without $(NH_4)_2SO_4$. After determination of $FA_{1/2}$ they were returned to the standard condition (pH 5.5 without $(NH_4)_2SO_4$) to check for column robustness, with 3–6 wash volumes before sampling. Data from Ref. [64].

Table 1. The effect of 100 mM $(NH_4)_2SO_4$ on Nd^{3+} recovery (percentage of 1 mM input metal removed) by 10–11 month old biofilms at pH 5.5 and 3.5.

of activity (~10%) seen after exposure to pH 3.5 and return to pH 5.5. Columns were returned to the original conditions after each use to check that the potential for column re-use was not compromised for potential re-use in multiple cycles.

7. An alternative approach for REE recovery via biogenic phosphate

Uptake of Ce(III) was reported by the Gram negative microorganism *Pseudomonas fluorescens* [65]. Ce(III) uptake was accompanied by its oxidation to Ce(IV). A 10-fold increase in the number of bacteria present reduced the proportion of Ce(IV) formed, from 80 to 50%, suggesting a mechanism of 'fixation' of the Ce(III) into a form on the cells which was resistant to oxidation. A longer contact time resulted in a greater amount of the 'fixed' Ce(III) following addition of MnO_2 as an oxidising agent. Cerium phosphate nano-particles were formed on the cells in the absence of added phosphate and hence the formation of Ce(III) phosphate at the expense of stored and mobilised phosphate was suggested [65]. Accordingly, storage of phosphate as intracellular polyphosphate bodies in *Pseudomonas putida* has been reported [66].

Ohnuki et al. [65] suggested two mechanisms of Ce(III) uptake, a rapid sorption followed by a slower Ce(III) PO_4 deposition. Analysis by time resolved laser fluorescence spectroscopy (TRLFS) that showed that the initial adsorption of Eu(III) on Gram negative *Ps. fluorescens* occurs through multi-dentate and inner-sphere complexes [6], while X-ray extended fine structure (EXAFS) analysis of Sm^{3+} uptake by Gram positive *Bacillus subtilis* showed that Sm^{3+} is complexed with carboxylate and phosphate functional groups [8].

The conclusion regarding a secondary, slower process of REE(III) fixation via phosphate deposition via stored reserves is in accordance with the earlier results. Jiang et al. [9] observed formation of Ce(III) and Yb(III) phosphates by the yeast *Saccharomyces cerevisiae* over 2 days. When lysed for 2 months after development, $YbPO_4$ nano-particles persisted in the solution showing the stability of the nano-material [9]. Jiang et al. [9] attributed this to the very low solubility product of REE phosphates (e.g. $CePO_4$ $\log k = -26.2$) as compared to, for example, uranium minerals (e.g. for autonite $\log k = -19.3$ [9]).

Up to date, apart from the *Serratia* immobilised cell system (above), there has been little attempt at bio-process development using bio-mineralisation systems. Delivery of the mineralising microorganisms into the target solution and separation of the harvested REE phosphate does not seem to have been developed apart from the use of immobilised biomass columns. It should be noted, however, that while the *Serratia* enzymatic mechanism was irreparably damaged at pH 3.5 and below [64], Yb phosphate mineralisation by *S. cerevisiae* occurred at pH 3 (although the longevity was not tested). EXAFS analysis of the deposited material showed a similar Yb(III)-edge for samples made at pH 4 and 5 while that for sample made at pH 3 was subtly different [67].

8. An alternative biotechnology for REE recovery: use of biogenic hydroxyapatite

Hydroxyapatite (HA), a major component of natural bone, has been used as a material for sequestering radionuclides using bone-HA [68]. However, use of animal by-products is subject to legislative constraints [69]. Using the same mechanism as described for phosphatase-mediated deposition of $NdPO_4$, (above) deposition of biogenic hydroxyapatite was achieved using biofilm-immobilised cells [70]; use of biofilm, as described above, yielded a highly adhesive bio-HA layer [71] which was tightly held onto the support after HA-mineralisation The work done to remove biofilm from a polypropylene support was measured (using a micromanipulator) at 3.4 ± 0.7 and 126.2 ± 23.3 J/m^2 for native and bio-mineralised biofilm, respectively [71].

The *Serratia* bio-HA was found to accumulate Eu^{3+} [72, 73]. This study focuses on the formation of cell-bound HA-crystals with the aim to produce smaller crystallites than those obtained by 'classical' methods which include batch growth of bacteria as used previously [74, 75]. Previous work on the bio-deposition of $NdPO_4$ by continuously pre-grown cells showed that, despite extensive bio-mineralisation, an average crystallite size of ~14 nm was obtained [21]; the cells apparently exert a stabilising effect that restricts 'runaway' crystallisation and nanoparticle agglomeration. One rationale for this is that, since continuously pre-grown cells produce copious extracellular polymers and since these polymers provide nucleation sites [76], cells pre-grown under continuous culture should produce more and smaller (for the same amount of Ca^{2+} provided) crystals than batch-grown cells, being stabilised against agglomeration by the surrounding extracellular polymeric material (EPM) matrix. A biofilm comprises cells bound together via copious EPM [68], the composition of which is dependent on the growth conditions [76, 77]. It is well known that Ca^{2+} ions crosslink adjacent EPM chains via their carboxyl groups and, in this respect, the bacterial EPM-Ca (subsequently EPM-HA) complex could be considered as a biogenic metal-organic framework (MOF), supplemented with ingressing phosphate via the co-localised phosphatase enzyme resulting in localised nano-precipitation under bio-control. Since bio-HA on *Serratia* was shown to take up Eu^{3+} (above), this prompted a study of the use of this material as a potential scavenging agent for REEs which does not require the use of living cells. Dried, heat-treated bio-HA accumulated metals as well as fresh (hydrated) bio-HA [73] give a bio-mineral with a long shelf-life and less encumbered by restrictions on use as compared to animal by-product materials.

The material deposited in the biofilm within the EPM (**Figure 9A**) and biofilm between the cells (**Figure 9B** and **D**) was confirmed to be hydroxyapatite using selected area X-ray diffraction and X-ray powder diffraction analysis (XRD) [78] and estimation of the bio-HA nanoparticle size by the XRD patterns under various incubation conditions gave an average of 20–25 nm [78, 79] as compared to the ~40 nm made by batch-grown cells [75] and ~80 nm of commercial hydroxyapatite (**Table 2**) [73].

Two features are highlighted by **Figure 9**. It is known that the bacteria metabolised the carbon components of G2P and citrate as provided in the mixture (analysis showed the disappearance of both citrate and glycerol P. Yong, M. Paterson-Beedle and L.E. Macaskie, unpublished); in the absence of any nitrogen source the cells did not grow but instead deposited intracellular deposits of the storage material polyβ-hydroxybutyrate [58] as shown by the electron

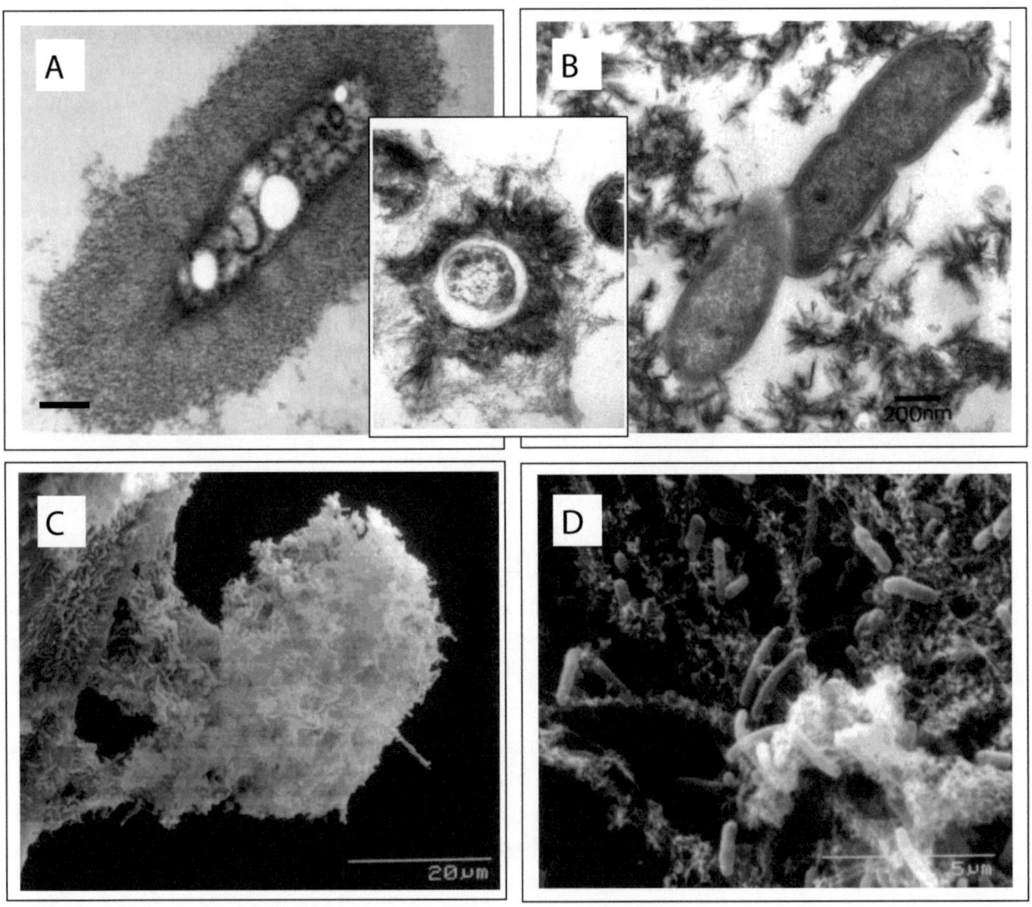

Figure 9. Formation of hydroxyapatite (HA) by *Serratia* cells. (A) Initial deposition of HA nano-crystallites within the EPM layer surrounding the cell. (B) As HA deposition proceeds (via daily dosing of Ca^{2+} and G2P) the material assumes needle-like deposits of HA extruding from the cells. Inset: Progression of the bio-HA (cross section) from cell-bound to needle-like crystals within layers of EPM Images: transmission electron microscopy; bars = 200 nm. (C) Mass of bio-HA from cut spur of polyurethane foam. (D) Top view of HA-biofilm on foam. Note extensive mineralised deposits cross linking the layer of cells (images: scanning electron microscopy).

Treatment (°C)	Organic content (%)	Surface area (m^2/g)	Crystallite size (nm)	Uptake of metal (mmol/100 g material)	
				Eu^{3+}	UO_2^{2+}
20	36	ND	40	154	131
200	27	65	34	154	131
250	23	76	37	154	131
300	16	93	31	154	131
350	9	102	37	154	131
400	7	115	36	154	131
450	3	86	36	154	131
500	3	61	32	154	131
550	1	38	48	154	131
600	2	27	64	154	131
650	0	17	125	153	124
700	0	12	271	151	116
Comm-HA	0	21	85	13	130

Data from Handley-Sidhu et al. [73]. Bio-HA was made by daily dosing with 2 mM Ca^{2+}, 5 mM G2P and 2 mM citrate buffer in 0.1 M AMPSO buffer, pH 8.6. Comm-HA: commercial HA (Sigma-Aldrich). Metals were used at 10 mM individually (aq; pH 5–6); no mixed metal tests were carried out. Eu uptake was 0.23 mg/mg bio-material (0.32 mg/mg HA component); U uptake was 0.31 mg/mg bio-material (0.42 mg/mg HA component) i.e. ~30 and 40% of the mass of the native bio-HA, respectively. Errors were within 5%.

Table 2. Uptake of Eu^{3+} and UO_2^{2+} into bio-HA pre-treated at increasing temperatures.

transparent inclusion bodies in **Figure 9A**; the degree of carbon turnover was not established. Secondly, the pattern of bio-HA deposition was shown as an initial formation of electron opaque material in the EPM around the cell (**Figure 9A**; capsular material, after 3 days of daily dosing with Ca^{2+} and G2P), progressing to intense mineral deposition within the EPM and formation of needle-like crystals that egressed around and beyond the cells and in the looser EPM in the regions between them, and densification to form a mineral network (**Figure 9B** and **D** and inset). Examination by scanning electron microscopy (a cut end of the foam is presented) shows the role of biofilm in adhering bio-HA to the support (**Figure 9C**) and in addition mineralised strands of EPM are visible between the cells and crosslinking them (**Figure 9D**). Many of the cells have become separated from the EPM-HA matrix leaving a meshwork of HA strands (**Figure 9D**)—a MOF that is typified by a crystalline material with voids to permit solute permeation. The biomaterial was evaluated for its ability to take up Eu^{3+}.

9. Eu^{3+} uptake by bio-HA made by cell suspensions

Early work used cells that had been pre-grown in batch culture suspensions (no carbon restriction). This reduces their ability to form biofilm (and hence EPM) [77] and hence continuously pre-grown and carbon-restricted cells were routinely adopted and bio-HA was made by daily dosing of re-suspended cells with a mixture of Ca^{2+}, G2P and citrate. Initial studies established that bio-HA made in the presence of citrate was more efficient than commercial HA in removing Sr^{2+} from an artificial groundwater [74] and hence citrate (generally 2 mM) was routinely incorporated into the Ca^{2+}/G2P challenge solution to make bio-HA. Following initial work by Holliday et al. [72] a subsequent study provided Eu^{3+} to a preparation comprising 26% organic material and hence 64% loading of HA (20 mg mass of bio-HA) (**Table 2**); the uptake of Eu^{3+} was ~30% of the HA dry mass (**Table 2**). Heating of the HA materials showed that bio-HA was little-affected even after removal of the organic component and potential agglomeration of the nano-crystallites and with a significant reduction in their surface area (**Table 2**). This suggested a mechanism other than simple surface adsorption and it was also noteworthy that while UO_2^{2+} was removed comparably by the bio- and commercial-HA, the latter showed negligible ability to remove Eu^{3+} which suggested differences in the nature of biogenic and chemical HA materials.

In a subsequent study (daily dosing with mixtures of $CaCl_2$, G2P and citrate at pH 7.0, 8.6 or 9.2: 8 days; all of the Ca^{2+} was removed each day) a known mass of bio-HA was made onto a known mass of bacteria (to give 50% loading as HA); here, the Eu^{3+} uptake reached ~42 mg/100 mg bio-HA or 0.84 mg mineral (i.e. 84% of the mineral mass) [78]. The HA-crystallite size was 21 ± 1.6 nm from materials made at pH 7 and 8.6 (pooled data) [79].

Comparing HA-bio-material made at pH 7, the optimal mixture was 2 mM Ca^{2+}, 2 mM citrate and 5 mM G2P [79] (5:2 of G2P:Ca^{2+}) and changing this ratio to 1:5 reduced the capacity for Eu^{3+} substantially (**Figure 10**).

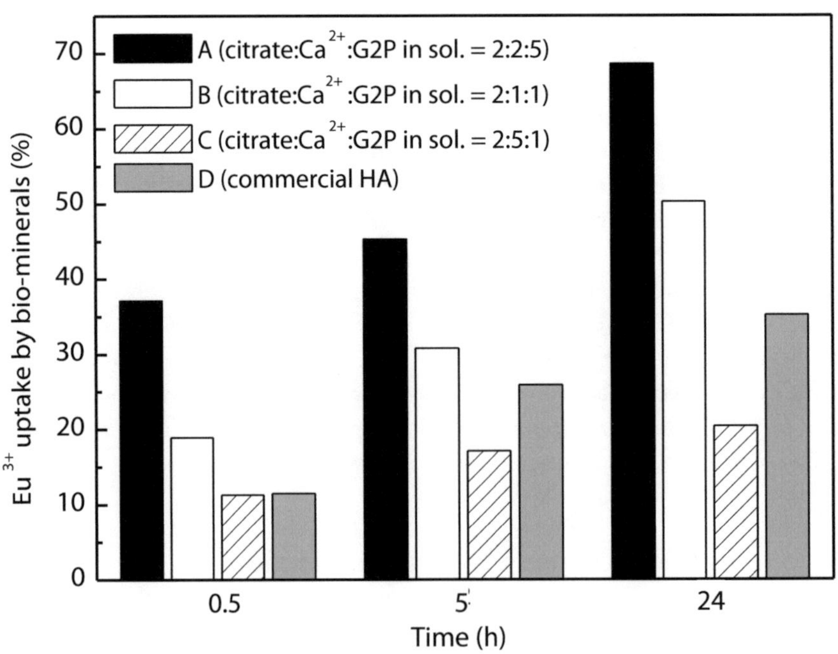

Figure 10. Uptake of Eu^{3+} made by bio-HA at pH 7 by methods A, B and C of bio-HA synthesis prior to Eu^{3+} challenge [79]. No further uptake occurred after 24 h solution (sol.) mixtures (citrate:Ca^{2+}:G2P) A: 2:2:5; B: 2:1:1; C: 2:5:1; D: commercial HA.

10. Eu^{3+} uptake by biofilm-HA immobilised in flow-through columns

The biofilm component of the continuously pre-grown bacterial culture was retained for use in this study (column tests). The planktonic cells fraction (from the vessel outflow) was used in previous work [78, 79]. Cells were from two independent cultures (I and II). Biofilm from the continuous culture vessels as immobilised biofilm-HA in a flow-through column. In order to be able to compare the results with a previous biofilm-HA adhesion study that quantified and confirmed the strength and durability of biofilm-HA [71], the biofilm was loaded with HA at pH 9.2. The HA-loading was calculated (by loss of Ca from the column exit solution) to be ~5 times the dry weight of bacteria (2300 mg of HA per column of eight biofilm discs). Following loading with HA and washing, the columns were challenged with 0.5 mM $Nd(NO_3)_2$ (aq) at 4 mL/h (0.5 column fluid vol/h). The exit solution was monitored and the Eu^{3+} breakthrough point was determined (**Figure 11**).

The bio-HA columns (but not the control) removed Eu^{3+} (~95%) during 2 L after which the activity decreased. The breakthrough region was ~2 L with a large data scatter after ~2.4 L. The column was not saturated since Eu^{3+} removal was sustained at ~50% removal until 4 L (mean value: **Table 3**) when the experiments were stopped. A mass balance analysis was calculated by averaging each point for experiments I and II and then calculating the mean of all samples between 2 and 4 L of flow. Initially (up to 600 mL) some phosphates was evolved which suggests a release of loosely bound HA material. From 600 mL to 2 L this was 0.1 mM which, subtracted from the Ca^{2+} released, indicates stoichiometry between Eu^{3+} removed and Ca^{2+} evolved and hence an ion exchange mechanism. The 1:1 stoichiometry was clearly seen in the second flow period (2–4 L) (**Table 3**).

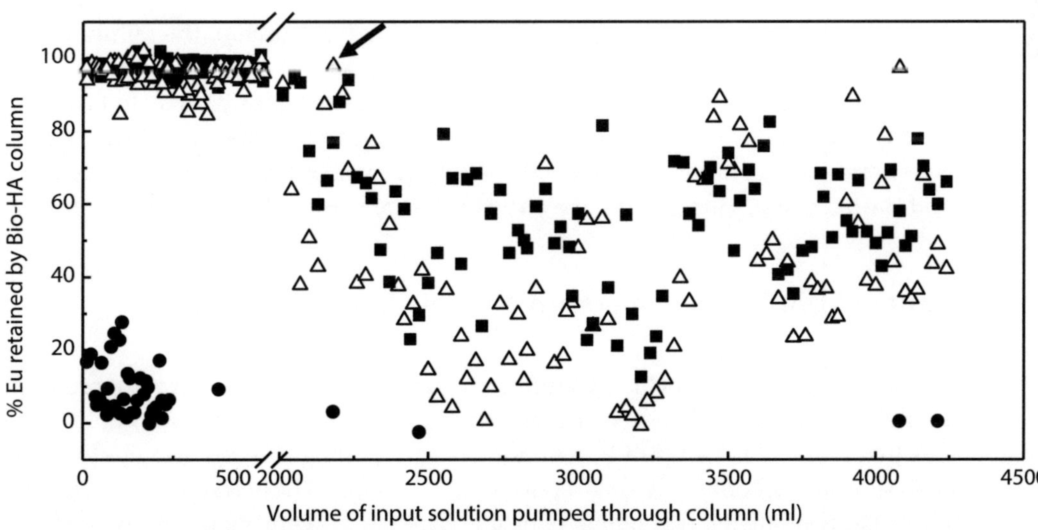

Figure 11. Uptake of Eu^{3+} by bacterial HA. The setup was shown in Figure 2 except that the loading solution was 1 mM Ca^{2+}, 5 mM glycerol 2-phosphate and 2 mM citrate buffered to pH 9.2 (25 mM TAPSO) for the loading of hydroxyapatite. The columns from independent biofilm preparations I and II (■, △) were drained, and the flow changed to 0.5 mM $Eu(NO_3)_3$ (aq). Column exit solution was assayed for Eu^{3+} colorimetrically; selected samples were retained for analysis of Eu, P and Ca by inductively coupled plasma optical emission spectroscopy (ICP-OES) in Table 3. ●: control (biofilm only). Arrow: Eu^{3+} breakthrough.

Volume passed (L)	Eu^{3+} accumulated (mmol/L)	Ca^{2+} evolved (mmol/L)	Pi evolved (mmol/L)
0–2	0.47 ± 0.04	0.56 ± 0.31	0.28 ± 0.02 (0-0.6 L)
			0.10 ± 0.02 (0.6–2 L)
2–4	0.22 ± 0.03	0.23 ± 0.01	0.017± 0.003 (2–4 L)

Table 3. Eu^{3+} uptake into immobilised bio-HA and Ca^{2+} and phosphate (Pi) evolved.

Since the column was not saturated with Eu^{3+} by the 4 L stage, this study suggests that a surface layer of bio-HA was initially saturated and then a second phase occurred whereby incoming Eu^{3+} exchanged for Ca^{2+} in deeper layers of the HA-biofilm in a slower process. The variation of Eu^{3+} removal in the second stage ranged from 0 to 100% which possibly reflects discontinuous access of new parts of the bio-HA as the column becomes saturated in the readily accessible areas. **Figure 9C** and **D** suggests that the bio-HA is not homogeneous in the biofilm. While magnetic resonance imaging (MRI) has been applied to visualise biofilms in the foam *in situ* and also during HA-loading [70] and could be used to monitor progressive ingress of the paramagnetic REE^{3+} species (Gd is a well-known contrast agent in MRI due to its paramagnetic effect) as described for mapping of paramagnetic Cr(VI) in columns [80]. The resolution of MRI (~100 μm) is insufficient to visualise even the detail of the foam struts and the biofilm appeared as an indistinct layer on them [70]; hence spatial analysis of the Eu^{3+} penetration into the bio-HA would be very difficult, additionally because of the long image acquisition times required. More extensive work is required to optimise the bio-HA loading on the biofilm and to obtain a more even distribution. It may also be found useful to use a fluidised bed delivery to ensure more even penetration throughout all exposed surfaces and to avoid channelling behaviour within the column.

Although bio-HA was found to take up both Eu^{3+} and UO_2^{2+} separately, the failure of the former in commercial HA only (**Table 2**) and the ability of the bio-material to take up Sr^{2+} from seawater (high saline) where commercial HA cannot [19] suggest bio-HA to be a new bio-agent for REE recovery. However, experiments using metal mixtures in realistic matrix backgrounds are needed in order to evaluate the scope for metal selectivity in actual applications. To aid further studies, insight into the structure of the bio-material is required.

11. Structure of bio-HA and mechanism of REE sequestration

The above discussions suggest that bio-HA differs in some way from its chemical counterpart. It is important to note that XRD analysis examines bulk sample and reports on only those components that are crystalline (even if in minor amounts), while selected area electron diffraction analysis examines only very small spots within a sample. The broad peaks for bio-HA [74, 78, 79] suggest the material contains poorly crystalline or amorphous components, as confirmed by the analysis of its crystallinity (below).

A detailed investigation was made of calcium phosphates made chemically in a groundwater matrix with and without added bacteria [81]. Ground waters contain the naturally occurring geopolymers, humic and fulvic acids and, with bacteria, organic acids derived from bacterial metabolism and also microbial extracellular polymers.

The mineral progressed from the initial formation of amorphous calcium phosphate and transformation to poorly crystalline HA within 7 days. The presence of bacteria delayed the onset of precipitation, resulted in changes to the lattice parameters and also reduced incorporation of trace elements as compared to cell-free systems. Importantly, that study [81] distinguishes between the ability of HA to form in a passive, bacterially supplemented system (with added external phosphate) [81], as compared to one in which the phosphate is being supplied enzymatically within the extracellular polymeric matrix [the *Serratia* example system described in this work]; both indicate that the matrix with/of the bacteria influences the formation of HA, its lattice structure and also its ability to take up secondary metals. A detailed discussion is outside the scope of this chapter (see [81]) but salient points highlighted were the effect of organic compounds with respect to HA crystallinity, also noting that uptake of Sr^{2+} into HA decreased in the presence of bacteria during HA synthesis (the uptake of Sr^{2+} into pre-formed HA was not tested). Specifically, it was cited that the presence of citrate (relative to acetate) in the mixture had the effect of decreasing the crystal size and showed both a higher content of impurities and a higher incorporation of carboxylate groups into the crystal structure [82]. We suggest that by fabricating the HA from within the EPM (**Figure 9A**), fed by enzymatically generated phosphate, a novel scaffolded HA is formed, with crystal growth controlled by the surrounding EPM-matrix, which may promote features enabling it to take up Eu^{3+} efficiently even after heating (i.e. a persistent matrix 'memory') whereas 'chemical HA' would lack this property (**Table 2**).

An early analysis of bio-HA made under the conditions described here (**Figures 10** and **11**; **Table 3**) confirmed formation of a hydroxyapatite material that converted to β-tricalcium phosphate (β-TCP) after sintering [83]. Further analysis [84] showed that non-sintered material comprised mainly calcium-deficient HA (CDHA) of ~9% crystallinity which increased to 53% after sintering at 1200°C, with the sintered material identified by X-ray diffraction and Fourier Transform Infrared spectroscopy as mainly CDHA with some sodium calcium phosphate. In the bio-HA (native, treated at 600°C and sintered at 1200°C) the stretching O-H bond (characteristic of the commercial HA spectrum at wavenumber 3566 cm^{-1}) was absent, suggesting that the HA crystal lattice is defective; Ledo et al. [84] discussed the possible presence of vacant calcium sites and hydroxide ion sites and/or that some of the phosphate ions may be either protonated or substituted by other ions. They also suggested that the EPM materials were associated with the bio-HA and were destroyed upon sintering. Thackray et al. [83] suggested that after sintering bio-HA comprised β-tricalcium phosphate as a dominant phase, together with small amounts of $Na_3Ca_6(PO_4)_5$ and $NaCaPO_4$. The Na^+ ions probably arise from the sodium salt of G2P used in bio-HA synthesis. They also noted that the presence of impurities may influence the phase composition, the final material after sintering reflecting the method used to produce it. Formation of CDHA or β-TCP after sintering depended on the initial Ca:P ratio, while crystals produced at pH 9.2 in the presence of citrate remained as HA while those produced at pH 8.6 without citrate sintered to β-TCP [84]. As bulk production of bio-HA would probably require heat treatment to remove both bacterial cells and pyrogens, these differences may impact upon the properties with respect to the subsequent REE uptake.

HA materials are polycrystalline with characteristic grain boundaries containing amorphous calcium phosphate species; the highly amorphous nature of bio-HA together with the high occurrence of organic materials (above) suggests that the occurrence of grain boundaries would be large but it is difficult to see how Eu^{3+} was loaded to >80% of the HA mineral weight [78] unless most of the material comprised ultra-small crystallites with a high proportion of surface calcium phosphate groups as compared to buried ones. It would be possible to calculate this in homogeneous material of known crystallite size but the nature of bio-HA and dependence on the exact conditions of synthesis would make this calculation very difficult.

In HA, 10 calcium atoms are aligned in two non-equivalent sites denoted as Ca(I) and Ca(II). It was reported that these sites are the target for divalent cation substitution [85] but the evidence for trivalent REE binding is less clear. Two studies had reported the incorporation of Eu^{3+} into the Ca(I) sites of apatite at room temperature via use of time-resolved laser fluorescence spectroscopy (TRLFS) (see [72]) but in contrast the trivalent actinide Cm^{3+} incorporated into the apatite mineral structure [72]. Holliday et al. [72] used TRLFS in conjunction with extended X-ray absorption fine structure (EXAFS) to eliminate the Ca(I) site as the target for Eu^{3+} incorporation, pointing out that poorly defined Eu^{3+} would exist in an amorphous environment with both the fluorescence lifetime and emission spectra indicating that Eu^{3+} does not incorporate into the HA crystal structure but with the excitation spectrum showing a broad peak that would be expected by location of Eu^{3+} in an amorphous environment [72]. Persistent biological residue was largely discounted since biologically produced and synthetic apatite gave identical fluorescence emission spectra; moreover, even after sintering to remove biological material only ~50% loss of crystallinity was achieved [83] i.e. a largely amorphous material with very small crystallites. Holliday et al. [72] concluded that Eu^{3+} incorporates into grain boundaries and not Ca(I) or Ca(II) sites.

β-TCP is a known feature of grain boundaries in bio-HA [83] (above) but TCP results from bio-HA after heat treatment whereas native material was used by Holliday et al. [72]. Hence, the data appear inconsistent, especially given the high loadings of Eu^{3+} observed into bio-HA in several independent studies [73, 78, 79] and since only a low uptake of Eu^{3+} was seen using commercial HA (**Table 2**), in accordance with a hypothesis that suggests non-penetration into the Ca(I) and Ca(II) sites of the crystalline material. Handley-Sidhu et al. [73] argued that removal of the organic material by heating left voids and hence a large number of grain boundaries for Eu^{3+} uptake and they also noted that there was no change in the XRD pattern after Eu-substitution, suggesting no incorporation into the mineral mass. However, XRD only provides information on the crystalline components, not the amorphous material, and, even after sintering, ~half of the bio-material remained amorphous [83].

The bi-phasic nature of the Eu^{3+} uptake into bio-HA (**Figure 11**), the high capacity (**Figure 10**) and the clear stoichiometry of Eu^{3+} uptake and Ca^{2+} release (**Table 3**) clearly argue for two mechanisms: an initial saturation of grain boundary sites (patterned by the original organic matrix) followed by a slower uptake via ion exchange. Inspection of three studies shows that Holliday et al. [72] used only 1 μM Eu^{3+} whereas Gangappa et al. [78] and Handley-Sidhu et al. [73] used 0.5 and 10 mM Eu^{3+}, respectively. It is likely that the Eu^{3+} in the former study [72] was sufficient only to bind (preferentially) to surface sites whereas a higher concentration of

Eu^{3+} allowed, over time, a bypass or migration of the surface-bound Eu^{3+} to deeper locations, resulting in a slower uptake phase with, presumably, migration of the surface located Eu^{3+} into deeper layers being rate-limiting. This was indirectly supported by the observation that it was difficult to describe metal uptake into bio-HA by classical Langmuir or Freundlich isotherm analysis [19]. If so, then the architecture of the surface of the HA is a key to the subsequent uptake and incorporation reactions giving >80% saturation of the HA. In a batch suspension this required ~24 h for completion [79] (**Figure 10**) whereas with the column study (using grams of HA material and a continuous Eu^{3+} feed: **Figure 11**) saturation of the first uptake phase occurred after 125 column volumes (250 h) but that of the second (ion exchange only: **Table 3**) was not seen even after 250 column volumes (500 h).

As described above, the presence of organic materials can impact upon the nano-crystalline format and maintains a large number of grain boundaries for the surface interactions but it is now becoming appreciated that organic ligands in the matrix (whether from the suspension or from the bio-matrix itself) also impact upon the actual surface structure (see earlier discussions). In parallel work, the presence of organic ligands from wastes was used in the background matrix during HA synthesis [18]. By this approach the uptake of Cu(II) into HA was increased by several-fold as compared to chemical HA made without chemical ligands, to 850 mg Cu/g HA (85% of the mass) [18]. However, when corrected for the different atomic masses of Cu and Eu the uptake of metal into the bio-HA (this study) was >2-fold higher than for the material made with applied ligands [18]. The extent to which the very high substitution of Eu^{3+} may destabilise the bio-HA lattice is not known but the lack of inorganic phosphate recovered (**Table 3**) suggests that the composite material is stable. A detailed study of its structure is underway.

12. Potential for selective REE recovery: applications of a magnetic field

Initial tests used the biofilm (enzymatic) system on columns challenged with either Nd^{3+} or Eu^{3+} with the columns positioned between external Nd permanent magnets in the enzymatic bio-mineralisation mode. The different magnetic properties of the REE [86] might be expected to enrich for the more magnetically active element(s) in the biofilm at the bottom of the column. However, with both Nd^{3+} and Eu^{3+} (separately) the columns rapidly lost activity. The reason for this was not investigated but it is likely that a metallic co-factor (e.g. Fe or Mn) was stripped from the enzyme; the role of Fe in the function of purple acid phosphatases (for example) is well known [87]. Hence, a magnetic separation would not be possible using the enzymatic recovery route and the ion exchange route into bio-HA is more attractive. Nd^{3+} and Eu^{3+} were loaded (separately) into bio-HA as described above and the material was squeezed from the biofilm-HA discs. While bio-HA alone gave no response, REE-substituted HA was collected rapidly onto the magnet [88]. The migration rates of Nd-HA and Eu-HA were not measured but the potential for magnetic separation using a bio-HA 'collector' in a magnetic field would be worthy of further exploration as a means for possible REE enrichment. A column system is non-ideal but future tests would envisage a fluidised bed held in a magnetic gradient; appropriate systems for separation of differentially magnetised-HA are under current investigation.

13. Potential for selective REE recovery: a biotechnology approach

Previous work has shown that pre-nucleation enhanced $NdPO_4$ deposition in aged biofilms [89]. This expanded from early work showing that pre-nucleation by one metal phosphate could promote the subsequent recovery of a different 'target' metal [29].

As part of the TBP study (above; S. Moriyama, I. Mikheenko and L.E. Macaskie, unpublished) (using 1 mM G2P/2 mM TBP and 1 mM metal ion in 2 mM citrate) the removal of uranyl ion by a column pre-nucleated with uranyl phosphate deposit (at the $FA_{1/2}$ value for removal of Nd^{3+} in the single metal flows) was less than 10%. Conversely, a column that had been pre-nucleated with $NdPO_4$ from the same background mixture gave enhanced removal of uranyl ion (35%) at the same flow rate. In contrast, pre-nucleation with uranyl phosphate followed by challenge with Nd^{3+} gave only 18% removal of Nd^{3+}, not 50% as in the single metal tests (although the pre-loading extents of $NdPO_4$ and HUO_2PO_4 were not the same). This suggests that, while a nucleation deposit of $NdPO_4$ may assist in the recovery of uranyl ion, the converse is true for neodymium recovery. This concept of a 'keying' (or indeed blocking) metal deposit laid down in advance of the REE mixture of interest suggested a possible development of this pre-nucleation concept towards selective recovery of REE. An initial survey [90] used suspended cells of 25 bacterial species expressing various levels of phosphatase activity. A selection of metals/bio-minerals was pre-deposited onto the cells prior to exposure to REE^{3+}. These initial metal deposits (low cost metals of low toxicity) were designated as 'Keying Metal A-Z.' With a high number of combinations, a column test in each case was impractical and these tests used batch suspensions. Selected examples are shown in **Figure 12**.

Figure 12A shows a strain pre-treated with three different 'keying' materials. Recovery of the REE was ~15–25% (mol/mol of the starting amount) but with little selectivity between them. Although Lu was rejected (i.e. enriched in the residual solution) negligible enhancement of REE removal was observed using three 'keying metals'. In contrast (**Figure 12B**) strain 17B when 'keyed' with material Z showed some selectivity. Native and 'keyed' cells rejected Lu^{3+} and Gd^{3+} (**Figure 12B**) and the 'keying' procedure increased the rejection of La^{3+} and Pr^{3+}. Selectivity was shown for the light REEs by up to 2-fold.

The rejection of La^{3+} and Pr^{3+} implies a competition between these and the 'keying' metal for deposition sites. Metal uptake in this set of experiments was not via phosphatase activity (no phosphate donor was provided) and a metal uptake mechanism akin to REE uptake into pre-mineralised cells (above) or via ion exchange (as with bio-HA) is implied or, indeed, metal diversion and uptake onto other bacterial surface sites following blockage of preferred in the 'keying' step. Analysis for keying metal appearing in solution and mass balances (e.g. as in **Table 2**) has not been attempted and would form the focus of future work.

Bio-sorption of metals onto functional groups of microbial cells is very well documented since the 1980s [91] and is now enjoying a resurgence as the price of metals rises (as well as environmental concerns about the impact of primary resource extraction), via a new generation of researchers (e.g. [92]). It is known which biological ligand group favours a particular

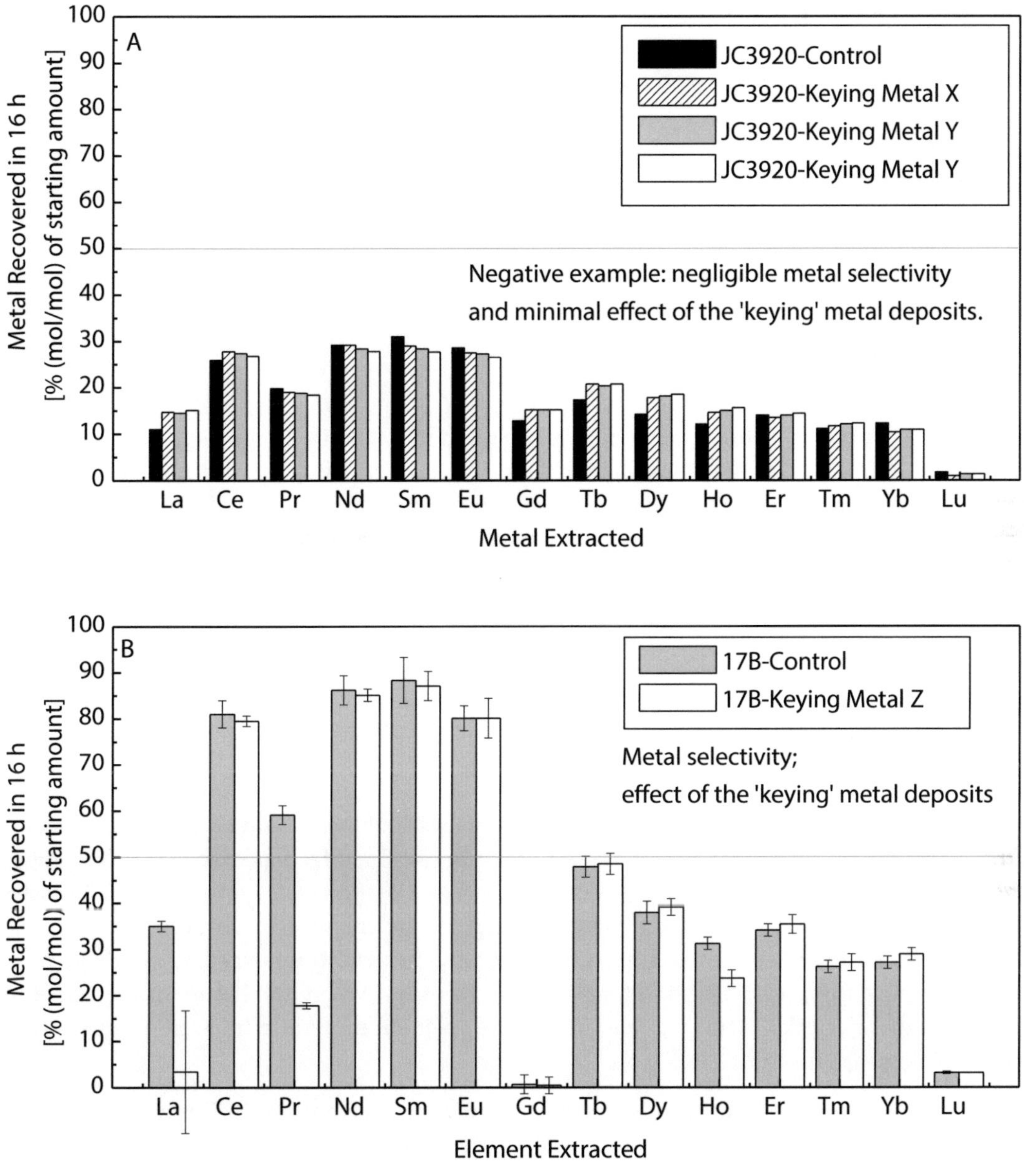

Figure 12. Effect of 'keying' bacteria with another metal deposit prior to exposure to metal of interest. In all experiments, equimolar amounts (approximately 1 mM in all cases, assessed by clean-room ICP-MS) of La-Lu as REE^{3+} chlorides were added as a mixture to bio-recovery medium (buffered to pH 7.2, $CaCl_2$, G2P and glycerol) from individual stock solutions immediately after supplementing with pre-grown, active cells of the bio-recovery organism under test (final volume 50 mL). After 16 h exposure at 22°C with gentle agitation, cells were harvested by filtration and washed thoroughly with deionised water to remove unbound metals before digestion in hot brominated *aqua regia* to degrade the biomass. Biomass-associated REE were analysed in the digestate by clean-room ICP-MS. Cells were also pre-treated with the 'keying' metals by a proprietary method. All data shown are expressed as % (mol/mol) recovery of metal from solution; error bars are standard error of the mean (N = 3). Reproduced from Ref. [90] with permission. (A) shows an example where there is no metal selectivity with keying deposit (strain JC920). (B) shows an example of metal selectivity using strain 17B with 'keying metal' deposit.

metal but the concept of metal-organic frameworks (MOFs; see earlier) with respect to the role of bacterial cell surfaces as potential MOFs is almost unexplored. In the current example (**Figure 12B**) it is possible that the second metal inserts into an extant MOF on the bacteria, which may be unique for each bacterial species (and also their specific growth conditions) and metal combination(s) as well as other ligands in the solution matrix, and the pH. The preliminary study made by Boden et al. (**Figure 12**) informs, by the use of known 'blocking', a future combinatorial chemistry and synthetic biology approach, given the large number of REE and relatively recent focus on their chemistry.

14. Future outlook

For process application the *Serratia* enzymatic method was initially proposed to provide a rapid and efficient way to achieve non-specific REE accumulation against uranyl ion into a mineral concentrate. However, this over-simplistic projection does not take into account the possible formation of metal organic frameworks in microbial EPM and the external solution in a bimetallic system that could 'push' a readily available metal into a less-available form. The use of TBP is shown to be of no benefit in the separation of REE and uranium, serving only to reduce metal recovery overall. Clearly an enzymatic route to metal recovery from real wastes may have limitations where metal separation is required, although this approach was highly effective in a uranium-only system [33, 47] and in the separation of REE(III) and Th(IV) by the restriction of flow residence time (**Figure 3**).

A study of HA formation by the same bacteria showed extensive deposition and, in this sense, the initial complexation of the Ca^{2+} into the bacterial exopolymer could be taken as an example of a simplistic, single metal MOF in the initial stages of bio-HA deposition. This concept of selectivity of metals into natural matrices to provide nucleation or 'keying' deposits to assist in the selective recovery of a second metal is a step change, inspired by observational studies, that now requires the concerted applications of combinatorial chemistry and synthetic biology to achieve further advances in targeted REE recovery, where the nature of the bacterial EPM or other bacterial surface layers provides a potential tool for manipulation using synthetic biology approaches.

Commercial implementation would assume no current selectivity of REE recovery and initially would aim to obtain a bulk metal phosphate bio-mineral for onward refining or prior to making a more concentrated leachate for a secondary selective REE recovery (possibly an ammonium sulphate leach, then a second, alkaline stage using bi-carbonate ion) but the overall economics is key. The possibility to provide a uranium concentrate as a high value side stream would be important. Within a previous EU consortium project [47], the cost of the *Serratia* bio-recovery process was estimated as comparable to ion exchange in terms of capital outlay but operational cost (the need for G2P for continued metal synthesis) was limiting and TBP is concluded to have little potential for realistic substitution for G2P due to complexities of the bio-chemistry of its utilisation. A similar conclusion regarding alkyl phosphate utility was reported elsewhere (see [81]).

In contrast to the high current commodity value of REE, uranium was of little importance in 1990s when the uranium recovery study was done [33, 47] (US$ 10/lb) but, with a global nuclear expansion, in response to concern about fossil fuel availability and the environmental consequences of CO_2 emissions, the economic and environmental impacts of a dual metal recovery become compelling.

With respect to the bacteria required for metal recovery, custom-growing of bacteria is costly. However, waste could be used, for example by pre-growing the bacteria on lactose (e.g. dairy waste) and ammonium phosphate (e.g. fertiliser) [47]. In terms of metal recovered against 'biomass elemental building blocks' (C, N, S, P), at a metal loading of 5–10 times the bacterial dry weight (reported for bio-mineralisation within columns) the 'building blocks' would account for a small proportion of the total solid upon ashing for delivery into current pyro- and hydro-metallurgies and bio-recovery represents a mechanism to convert large volumes of dilute metals into solid concentrates.

The high value of REE, and difficulty of their extraction by conventional methods, suggests a potentially cost-effective process, especially if the 'bulk recovery' *Serratia* column can be re-used. A secondary selective step from leachates from the *Serratia* column will require future synthetic biology developments for bio-material engineering towards REE selectivity but the feasibility is shown in **Figure 12B** and below.

To overcome the economic bottleneck of bulk phosphate supply, phytic acid, a plant storage material (a waste from bio-diesel production from plant materials), was shown to be cleaved enzymatically to support metal removal [93] although the low water-solubility of phytic acid (inositol phosphate) may pose operational difficulties.

In a worst-case scenario, where REE recovery from the *Serratia* column via leaching into a concentrate for a selective secondary REE recovery is not achieved, the fall-back option would be delivery of a concentrated solid into commercial refining. Since a large proportion of the cost of ore and waste solids processing involves comminution (a large energy demand and hence high environmental CO_2 burden) microbial generation of nano-phase concentrated materials for recycling would represent energy savings (less comminution of finely divided material) as well as waste volume reductions and hence transportation and handling costs. This process intensification (as well as various Life Cycle Analyses) is the current status of the biotechnology approach to REE bio-refining, and this chapter highlights the future potential and currently identified bottlenecks, as well as the large scope for both combinatorial chemistry and synthetic biology as combined tools. In this respect, 'biosorption' of metals has 'come of age' in that a vast literature of biosorption studies exists over the past 50 years for many types of microorganism, bio-polymers and metals/metal combinations and the use of data mining and *in silico* methods may prove essential to inform and shorten the delivery of combinatorial chemistry and synthetic biology approaches.

Following early studies on the use of selective metal binding peptides and proteins on cells [94, 95], two recent studies highlight this potential. An acidophilic fungal isolate (a *Penidella* sp.) accumulated dysprosium and other REEs to 0.9 mg/g dry biomass (9% of the biomass dry weight) [96], while a strain of the stalked, biofilm-forming bacterium *Caulobacter crescentus* was engineered to display lanthanide binding tags on the cell surface, endowing specificity for some REEs over others and selectivity against Ca^{2+} [4]. The 'tags' were attached to the protein S-layer surrounding the cells. The sorption of Tb^{3+} was stated [4] as 50 μM at a cell concentration of 8×10^8 cells/mL (i.e. assuming the mass of a single cell as 1 pg, an uptake of approximately 1% of the bacterial dry weight, comparing favourably with other biosorption systems) but it was not stated whether the cells were saturated with Tb. A low capacity is acceptable if the sorbent can be re-used; the REEs could be eluted using citrate in sequential sorption/elution cycles [4]. A mixed REE test solution was made from an *aqua regia* extract of an environmental sample (125°C; 8 h), hence the background organic matrix would have been modified before use during tests. Thermodynamic modelling was done according to established models (from 2002 to 2006 [4]) for a solution of Tb, Ce, citrates and acetates to assess the likely speciation of Tb^{3+} but the potential contributions of additional metals and, indeed, potential MOFs in real samples, would be important to establish in progression to real life application.

Acknowledgements

The authors acknowledge with thanks the support of NERC (Grant NE/L014076/1) and EPSRC (EP/M012719/1 supporting joint UK-Japan civil nuclear research) and the collaboration of The University of Plymouth, UK: Dr Rich Boden who obtained the data shown in **Figure 12** (taken from [90] with permission; A.J. Murray et al., senior author L.E. Macaskie; printed in 'Proceedings COM 2014 International Conference, Hyatt Regency Hotel, Vancouver Sept 28-Oct 1, 2014'). They wish to thank Drs R. Clough for help in ICP-OES analysis and S. Ward, without whom essential data could not have been obtained. **Figure 4** is reproduced from [16] with permission from the Royal Society of Chemistry, while **Figure 7** has been re-drawn from data presented in reference [57]. We also thank Drs P. Yong and M. Paterson-Beedle for use of unpublished information and Dr J.A. Hriljac (School of Chemistry, University of Birmingham) for helpful discussions.

Author details

Lynne E. Macaskie*, Sayo Moriyama, Iryna Mikheenko, Sarah Singh and Angela J. Murray

*Address all correspondence to: l.e.macaskie@bham.ac.uk

School of Biosciences, University of Birmingham, Edgbaston, Birmingham, UK

References

[1] Anon. Development of a Sustainable Exploitation Scheme for Europe's Rare Earth Deposits. http://www.eurare.eu. [Accessed 30-01-2017]

[2] Sun X, Waters KE. Development of industrial extractants into functional ionic liquids for environmentally friendly rare earth element separation. ACS Sustainable Chemistry and Engineering. 2014;**2**:1910-1917. DOI: 10.1021/sc500255n

[3] Zhang MM, Li HX, Andrade M, Xuan W, Correa A. Feasible bioprocessing technologies for low grade iron ores al. Minerals and Metallurgical Processing Journal. 2015;**32**:78-87

[4] Park DM, Reed DW, Yung M, Elsamimanesh A, Lencka MM, Anderko A, Fujita Y, Riman RE, Navrotsky A, Jiao Y. Bioadsorption of rare earth elements through cell surface display of lanthanide binding tags. Environmental Science and Technology. 2016;**50**:2735-2742. DOI: 10.1021/acs.est.5b06129

[5] Moriwaki H, Yamamoto H. Interactions of microorganisms with rare earth ions and their utilization for separation and environmental technology. Applied Microbiology and Biotechnology. 2013;**97**:1-8. DOI: 10.1007/s00253-012-4519-9

[6] Ozaki T, Suzuki Y, Yoshida T, Ohnuki T, Nankawa T, Francis AJ. Association of Eu(III) with bacteria and organic ligands. Journal of Nuclear and Radiochemical Science. 2005;**6**:73-76

[7] Takahashi T, Chatellier X, Hattori KH, Kato K, Fortin D. Adsorption of rare earth elements onto bacterial cell walls and its implications for rare earth sorption onto natural microbial mats. Chemical Geology. 2005;**219**:53-67. DOI: 10.1016/j.chemgeo.2005.02.009

[8] Takahashi Y, Yamamoto M, Yamamoto Y, Tanaka K. EXAFS study on the cause of enrichment of heavy rare earths on bacterial cell surfaces. Geochimica et Cosmochimica Acta. 2010;**74**:5443-5462. DOI: 10.1016/j.gca.2010.07.001

[9] Jiang M, Ohnuki T, Tanaka K, Kozai N, Kamiishi K, Utsunomiya S. Post-adsorption process of Yb phosphate nanoparticle formation by *Saccharomyces cerevisiae*. Geochimica et Cosmochimica Acta. 2012;**93**:30-46. DOI: 10.1016/j.gca.2012.06.016

[10] Gadd GM. Metals, minerals and microbes: Geomicrobiology and bioremediation. Microbiology. 2010;**156**:609-643. DOI: 10.1099/mic.0.037143-0

[11] Van der Watt JG, Wanders FB. Leaching of rare earth elements from bentonite clay. Journal of South African Institute of Mining and Metallurgy. 2012;**112**:273-285

[12] Goyne KW, Brantley SL, Chorver J. Rare earth element release from phosphate minerals in the presence of organic acids. Chemical Geology. 2010;**278**:1-14. DOI: 10.1016/j.chemgeo.2010.03.011

[13] Shan X, Lian J, Wen B. Effect of organic acids on adsorption and desorption of rare earth elements. Chemosphere. 2002;**47**:701-710. DOI: 10.1016/S0045-6535(02)00032-2

[14] Nockermann P, Thijs B, Parac-Vogt TN, Hecke KV, Meervelt LV. Tinant B, Hartenbach I, Schleid T, Nguyen MT, Binnemans K. Carboxyl-functionalized task-specific ionic liquids for solubilizing metal oxides. Inorganic Chemistry. 2008;**47**:9987-9999. DOI: 10.1021/ic801213z

[15] Mann S. Biomineralization: Principles and Concepts in Bioinorganic Materials Chemistry. Oxford: Oxford University Press; 2001. pp. 210

[16] Nelson YM, Lion LW, Ghiorse WC, Shuler ML. Production of biogenic manganese oxides by *Leptothrix discophora* SS-1 in a chemically defined medium and evaluation of their Pb adsorption characteristics. Applied and Environmental Microbiology. 1999;**65**:175-180

[17] Nelson YM, Lion LW, Shuker ML, Ghiorse WC. Effect of oxide formation mechanisms on lead adsorption by biogenic manganese (hydr) oxides and their mixtures. Environmental Science and Technology. 2002;**36**:421-425. DOI: 10.1021/es010907c

[18] Mercardo DF, Magnacca G, Malandrino M, Rubert AA, Montoneri E, Celi L, Prevot AB. Gonzalez MC. Paramagnetic iron-doped hydroxyapatite nanoparticles with improved metal sorption properties. A Bio-organic substrates-mediated synthesis. ACS Applied Materials and Interfaces. 2014;**6**:3937-3946. DOI: 10.1021/am405217j

[19] Handley-Sidhu S, Mullan S, Grail Q, Albadarneh M, Ohnuki T, Macaskie LE. Influence of pH, competitor ions and salinity on the sorption of strontium and cobalt onto biogenic hydroxyapatite. Scientific Reports. 2016. Article No 23361. DOI: 10.1038/srep23361

[20] Tolley MR, Strachan LF, Macaskie LE. Lanthanum accumulation from acidic solutions using a *Citrobacter* sp. immobilized in a flow through bioreactor. Journal of Industrial Microbiology. 1995;**14**:271-280. DOI: 10.1007/BF01569939

[21] Murray AJ, Singh S, Vavlekas D, Tolley MR, Macaskie LE. Continuous biocatalytic recovery of neodymium and europium. RSC Advances. 2015;**5**:8496-8506. DOI: 10.1039/C4RA14892D

[22] Murray AJ, Singh S, Vavlekas D, Tolley MR, Macaskie LE. Biorecovery of rare earth elements: potential application for mine water remediation. Advanced Materials Research. 2015;**1130**:543-547. DOI: 10.4028/www.scientific.net/AMR.1130.543

[23] Friebe N, Nuyken O, Obrecht W. Neodymium-based Ziegler-Natta catalysts and their application in diene polymerization In: Nuyken O, editor. Neodymium-based Zeigler Catalysts Fundamental Chemistry. Berlin: Springer-Verlag; 2006. pp. 1-154. DOI: 10.1007/12_094

[24] Chen F, Zhu Y-J, Zhang K-H, Wu J, Wang K-W, Tang Q-L, Mo X-M. Europium-doped amorphous calcium phosphate porous nanospheres: Preparation and application as luminescent drug carriers. Nanoscale Research Letters. 2011;**6**:67-76. DOI: 10.1186/1556-276X-6-67

[25] Pisarska J, Soltys M, Zur L, Pisarski WA, Jayasankar CK. Excitation and luminescence of rare earth-doped lead phosphate glasses. Applied Physics B. 2014;**116**:837-845. DOI: 10.1007/s00340-014-5770-9

[26] Macaskie LE, Mikheenko IP, Yong P, Deplanche K, Murray AJ, Paterson-Beedle M, Coker VS, Pearce CI, Cutting R, Pattrick RAD, Vaughan D, Van Der Laan G, Lloyd JR. Today's wastes tomorrow's materials for environmental protection. In: Moo-Young M,

Butler M, Webb C, Moreira A, Grodzinski B, Cui ZF, Agathos S, editors. Comprehensive Biotechnology. Amsterdam Elsevier. 2011;**6**:719-725

[27] Jeong BC, Hawes C, Bonthrone KM, Macaskie LE. Localisation of enzymically enhanced heavy metal accumulation by a *Citrobacter* sp. and metal accumulation in vitro by liposomes containing entrapped enzyme. Microbiology. 1997:**143**:2497-2507. DOI: 10.1099/00221287-143-7-2497

[28] Macaskie LE, Jeong BC, Tolley MR. Enzymatically accelerated biomineralization of heavy metals-application to the removal of americium and plutonium from aqueous flows. FEMS Microbiology Reviews. 1994;**14**:351-367. DOI: 10.1111/j.1574-6976.1994.tb00109.x

[29] Macaskie LE, Lloyd JR, Thomas RAP, Tolley MR. The use of microorganisms for the remediation of solutions contaminated with actinide elements, other radionuclides and organic contaminants generated by nuclear fuel cycle activities. Nuclear Energy. 1996;**35**:257-271

[30] Macaskie LE, Empson RM, Lin F, Tolley MR. Enzymatically-mediated uranium accumulation and uranium recovery using a *Citrobacter* sp. Immobilized as a biofilm within a plug-flow reactor. Journal of Chemical Technology and Biotechnology. 1995;**63**:1-16. DOI: 10.1002/jctb.280630102

[31] Petit LD, Powell KJ. Stability Constants Database. UK: SC-Database IUPAC and Academic Software; 2008

[32] Jeong BC, Kim HW, Macaskie LE. Phosphotransferase activity of acid phosphatases of a *Citrobacter* sp. Federation of European Microbiological Societies Microbiology Letters.1997;**147**:103-108. DOI: 10.1111/j.1574-6968.1997.tb10227.x

[33] Macaskie LE, Yong P, Doyle TC, Manzano TM, Roig M. Bioremediation of uranium-waste: biochemical and chemical factors affecting bioprocess application. Biotechnology and Bioengineering. 1997;**53**:100-109. DOI: 10.1002/(SICI)1097-0290(19970105)53:1<100::AID-BIT13>3.0.CO;2

[34] Yong P, Macaskie LE. The role of sulfate as a competitive inhibitor of enzymatically mediated heavy metal uptake by *Citrobacter* sp: Implications in the bioremediation of acid mine drainage water using biogenic phosphate precipitant. Journal of Chemical Technology and Biotechnology. 1999;**74**:1149-1156. DOI: 10.1002/(SICI)1097-4660(199912)74:12<1149::AID-JCTB164>3.0.CO;2-0

[35] Yong P, Macaskie LE. The effect of substrate concentration and nitrate inhibition on phosphate release and heavy metal removal by a *Citrobacter* sp. Biotechnology and Bioengineering, 1997;**55**:821-830. DOI: 10.1002/(SICI)1097-0290(19970920)55:6<821::AID-BIT1>3.0.CO;2-I

[36] Vavlekas DA. Construction and evaluation of a modular biofilm-forming chamber for microbial recovery of neodymium and semi-continuous biofilm preparation. Tolerance of *Serratia* sp.N14 on acidic conditions and neutralized *aqua regia*. Environmental Technology. 2016;**38**:1-18. DOI: 10.1080/09593330.2016.1189971

[37] Jeong BC. Studies on the atypical phosphatase of a metal accumulating *Citrobacter* sp. [D.Phil Thesis]. UK University of Oxford; 1992.

[38] Jambor JL, Roberts AC, Owens DR, Grice JD. Zajacite-(Ce) a new rare earth fluoride from the Strange Lake deposit, Quebec, Labrador. Canadian Mineralogist. 1996;**34**:1299-1304. http://rruff.info/doclib/cm/vol34/CM34_1299.pdf

[39] Anon (May, 2016) http://www.marketwired.com/press-release/pele-mountain-provides-update-on-development-plans-and-progress-in-elliot-lake-tsx-venture-gem-2121413.htm [Accessed 10-12-2016]

[40] Tolley MR. The Biological Treatment of Liquid Wastes Containing heavy metals [D. Phil Thesis]. UK University of Oxford; 1993.

[41] Paterson-Beedle M, Macaskie LE, Lee CH, Hriljac JA, Jee KY, Kim WH. Utilization of hydrogen uranyl phosphate based ion exchanger supported on biofilm for the removal of cobalt, strontium and caesium from aqueous solutions. Hydrometallurgy. 2006;**83**:141-145. DOI: 10.1016/j.hydromet.2006.03.020

[42] Ohnuki T, Kozai N, Sakamoto F, Suzuki Y, Yoshida T. Biological change of chemical states of actinides and lanthanides—effects of organic acids. Energy Procedia. 2013;**39**:175-183. DOI: 10.1016/j.egypro.2013.07.204

[43] Amico P, Daniele PG, Ostacoli G, Sammartano S. Mixed metal complexes in solution Part 4. Formation and stability of heterobinuclear complexes of cadmium II-citrate with some bivalent metal ions in aqueous solution. Transition Metal Chemistry. 1985;**10**:11-14. DOI: 10.1007/BF00620623

[44] Raju U, Warker SG, Kumar A. Green synthesis of multi-metal-citrate complexes and their characterization. Journal of Molecular Structure. 2017:**1133**:90-94. DOI: 10.1016/j.molstruc.2016.11.084

[45] Öhström L. A short review of the terms 'Metal Organic Frameworks' The IUPAC project 'Coordination Polymers and Metal Organic Frameworks; Terminology and Nomenclature Guidelines [internet]. 2017. https://www.iupac.org/fileadmin/user_upload/databases/2009-012-2-200-short.review.pdf. [Accessed 2017-01-10] 2017.

[46] Freiser H, Morrison GH. Solvent extraction in radiochemical separations. Annual Review of Nuclear Science. 1959;**9**:221-255. DOI: 10.1146/annurev.ns.09.120159.001253

[47] Roig M, Kennedy JF, Macaskie LE. Biological rehabilitation of metal-bearing waste water. 1995. Final Report, EU Contract EV5V-CT93-0251. Brussels, The European Commission

[48] Davis Jr W. Solubilities of uranyl and iron (III) dibutyl and monobutyl phosphates in TBP solvent extraction systems. Contract No W-7405-eng 26 Oak Ridge National Laboratory, US Atomic Energy Commission[internet]. 1961. http://web.ornl.gov/info/reports/1961/3445600497148.pdf. [Accessed 29-01-2017]

[49] Rufus AL, Sathyaseelan VS, Narasimhan S, Velmurugan S. Dissolution of synthetic uranyl dibutyl phosphate deposits in oxidising and reducing chemical formulations. Journal of Hazardous Materials. 2013;**254-255**:263-269. DOI: 10.1016/j.jhazmat.2013.03.050

[50] Thomas RAP, Macaskie LE. Biodegradation of tributyl phosphate by naturally occurring microbial isolates and coupling to the removal of uranium from aqueous solution. Environmental Science and Technology. 1996;**30**:2371-2375. DOI: 10.1021/es950861l

[51] Owen S, Jeong BC, Poole PS, Macaskie LE. Tributyl phosphate degradation by immobilised cells of a *Citrobacter* sp. Applied Biotechnology and Biochemistry. 1992;**34**:693-707. DOI: 10.1007/BF02920590

[52] Thomas RAP, Macaskie LE. The effect of growth conditions on the biodegradation of tributyl phosphate and potential for the remediation of acid mine drainage waters by a naturally occurring mixed microbial culture. Applied Microbiology and Biotechnology. 1998;**49**:202-209

[53] Berne C, Montjarret B, Guountti Y, Garcia D. Tributyl phosphate degradation by *Serratia odorifera*. Biotechnology Letters. 2002;**26**:681-686. DOI: 10.1023/B:BILE.0000023030.69207.c0

[54] Berne C, Pignol D, Lavergre J, Garcia D. CYP201A2, a cytochrome p450 from *Rhodopseudomonas palustris*, plays a key role in the biodegradation of tributyl phosphate. Applied Microbiology and Biotechnology. 2007;**77**:135-144. DOI: 10.1007/s00253-007-1140-4

[55] Mennan C. Bacterial acid phosphatase and its application to waste remediation and metal recovery [PhD thesis]. University of Birmingham UK; 2011. http://etheses.bham.ac.uk/1017/1/Mennan11PhD1.pdf

[56] Tanaka N, Dumay V, Liao Q, Lange AJ, Wever R. Bromoperoxidase activity of vanadate-substituted acid phosphatases from *Shigella flexneri* and *Salmonella enterica* ser. Typhimurium. European Journal of Biochemistry 2002;**269**:2162-2167. DOI: 10.1046/j.1432-1033.2002.02871.x

[57] Michel LJ, Macaskie LE, Dean ACR. Cadmium accumulation by immobilized cells of *Citrobacter* sp. using various phosphate donors. Biotechnology and Bioengineering. 1986;**28**:1358-1365. DOI: 10.1002/bit.260280910

[58] Lugg H, Sannons RL, Marquis PM, Hewitt CJ, Paterson-Beedle M, Redwood MD, Stamboulis A, Kashani M, Jenkins M, Macaskie LE. Polyhydroxybutyrate accumulation by a *Serratia* sp. Biotechnology Letters. 2008;**30**:481-491. DOI: 10.1007/s10529-007-9561-9

[59] Wu CH, Lam BR, Chou J, Bill M, Brodie EL, Anderson GL, Hazen TC, Conrad ME, Henriksen J, Wright KE, Fujita Y. Microbial metabolism of triethylphosphate, a potential phosphate source for radionuclide mineralization processes. Eos Transactions American Geophysical Union Fall Meeting Supplement. 2009;**90**:H34A–08

[60] Bassil NM, Bryan N, Lloyd JR. Microbial degradation of isosaccharinic acid at high pH. International Society for Microbial Ecology Journal. 2014;**9**:310-320. DOI: 10.1038/ismej.2014.125

[61] Tian J, Yin J, Chi R, Rao G, Jiang M, Ouyang K. Kinetics on leaching rare earth from the weathered crust elution-deposited rare earth ores with ammonium sulfate solution. Hydrometallurgy. 2010;**101**:166-170. DOI: 10.1016/j.hydromet.2010.01.001

[62] Tian J, Yin J, Chi R, Rao G, Jiang M, Ouyang K. Optimisation of mass transfer in column elution of rare earths from low grade weathered crust elution-deposited rare earth ore. Hydrometallurgy. 2010;**103**:211-214. DOI: 10.1016/j.hydromet.2010.04.003

[63] Moldoveanu GA, Papangelakis VG. Recovery of rare earth elements adsorbed on clay minerals. II: leaching with ammonium sulfate. Hydrometallurgy. 2013;**131-132**:158-166. DOI: 10.1016/j.hydromet.2012.10.011

[64] Vavlekas D. Microbial recovery of rare earth elements from metallic wastes and scrap. [MRes Thesis]. University of Birmingham UK; 2016.

[65] Ohnuki T, Jiang M, Sakamoto F, Kozai N. Yamasaki S, Yu Q, Tanaka K, Utsunomiya S, Xia X, Yang K, He J. Sorption of trivalent cerium by a mixture of microbial cells and manganese oxides: Effect of microbial cells on the oxidation of trivalent cerium. Geochim et Cosmochim Acta. 2015;**163**:1-13. DOI: 10.1016/j.gca.2015.04.043

[66] Tobin KM, McGrath JW, Mullan A, Quinn J, O'Connor KE. Polyphosphate accumulation by *Pseudomonas putida* CA-3 and other medium chain length polyhydroxyalkanoate-accumulating bacteria under anaerobic growth conditions. Applied and Environmental Microbiology. 2007;**72**:1383-1387. DOI: 10.1128/AEM.02007-06

[67] Jiang M, Ohnuki T, Kozai N, Tanaka K, Suzuki Y, Sakamoto F, Kamiishi E, Utsunomiya S. Biological nano-mineralization of cerium phosphates by *Saccharomyces cerevisiae*. Chemical Geology. 2010;**277**:61-69. DOI: 10.1016/j.chemgeo.2010.07.010

[68] Dimović SD, Smičiklas ID, Šljivić-Ivanović MZ, Plećaš IB, Slavković-Beskoški L. The effect of process parameters on kinetics and mechanisms of Co^{2+} removal by bone char. Journal of Environmental Science and Health. A Toxic/Hazardous Substances and Environmental Engineering. 2011;**46**:1558-1569. DOI: 10.1080/10934529.2011.609454

[69] Anon. Guidance for the animal by-product industry. Department for Environment, Food and Rural Affairs, UK. 2014. https://www.gov.uk/government/collections/guidance-for-the-animal-by-product-industry. [Accessed 04-01-2017]

[70] Macaskie LE, Yong P, Thackray AC, Paterson-Beedle M, Marquis PM, Sammons RL, Nott KP, Hall LD. A novel non line of sight method for coating hydroxyapatite onto the surfaces of support materials by biomineralization. Journal of Biotechnology. 2005;**118**:187-200. DOI: 10.1016/j.jbiotec.2005.03.006

[71] Yong P, Paterson-Beedle M, Liu W, Zhang Z, Macaskie LE. A study of biofilm and non line of sight bio-hydroxyapatite coatings using a *Serratia* sp. Advanced Materials Research. 2009;**71-73**:741-744. DOI: 10.4028/www.scientific.net/AMR.71-73.741

[72] Holliday K, Handley-Sidhu S, Dardenne K, Renshaw J, Macaskie LE, Walther C, Stumpf T. A new incorporation mechanism for trivalent actinides into bioapatite: A TRLFS and EXAFS study. Langmuir. 2012;**28**:3845-3851. DOI: 10.1021/la300014a

[73] Handley-Sidhu S, Hriljac JA, Cuthbert MO, Renshaw JC, Pattrick RAD, Charnock JM, Stolpe B, Lead, JR, Macaskie LE. Bacterially-produced calcium phosphate nanobiominerals:

Sorption capacity, site preferences and stability of captured radionuclides. Environmental Science and Technology. 2014;**48**:6891-6898. DOI: 10.1021/es500734n

[74] Handley-Sidhu S, Renshaw JC, Yong P, Macaskie LE. Nanocrystalline hydroxyapatite biomineral for treatment of strontium from aqueous solutions. Biotechnology Letters. 2011;**33**:79-87. DOI: 10.1007/s10529-010-0391-9

[75] Handley-Sidhu S, Renshaw JC, Moriyama S, Stolpe B, Mennen C, Bagheriasi S, Yong P, Stamboulis A, Paterson-Beedle M, Sasaki K, Pattrick RAD, Lead JR, Macaskie LE. Uptake of Sr^{2+} and Co^{2+} into biogenic hydroxyapatite: Implications for biomineral ion exchange synthesis. Environmental Science and Technology. 2011;**45**:6985-6990. DOI: 10.1021/es2015132

[76] Bonthrone KM, Quarmby J, Hewitt CJ, Allan VJM, Paterson-Beedle M., Kennedy JF, Macaskie LE. The effect of the growth medium on the composition and metal binding behaviour of the extracellular polymeric material of a metal accumulating *Citrobacter* sp. Environmental Technology. 2000;**21**:123-134. DOI: 10.1080/09593330.2000.9618893

[77] Allan AJM, Callow ME, Macaskie LE. Effect of nutrient limitation on biofilm formation and phosphatase activity of a *Citrobacter* sp. Microbiology. 2002;**148**:277-288. DOI: 10.1099/00221287-148-1-277

[78] Gangappa R, Yong P, Singh S, Murray AJ, Macaskie LE. Hydroxyapatite biosynthesis by a *Serratia* sp and application of nanoscale bio-HA in the recovery of strontium and europium. Geomicrobiology Journal. 2016;**33**:267-273. DOI: 10.1080/01490451.2015.1067657

[79] Gangappa R, Farrier A, Macaskie LE. Eu^{3+} sequestration by biogenic nano-hydroxyapatite synthesized at neutral and alkaline pH. Geomicrobiology Journal. In press. DOI 10.1080/01490451.2016.1261966

[80] Beauregard D, Yong P, Macaskie LE, Johns ML. Using non-invasive magnetic resonance imaging to assess the reduction of Cr(VI) using a bio-palladium catalyst. Biotechnology and Bioengineering. 2010;**107**:11-20. DOI: 10.1002/bit.22791

[81] Wright KE, Hartmann T, Fujita Y. Inducing mineral precipitation in groundwater by addition of phosphate. Geochemical Transactions. 2011;**12**:8-20. DOI: 10.1186/1467-4866-12-8

[82] Van der Houwen JAM, Cressey G, Cressey BA, Valsami-Jones E. The effect of organic ligands on the crystallinity of calcium phosphate. Journal of Crystal Growth. 2003;**249**:572-583. DOI: 10.1016/S0022-0248(02)02227-3

[83] Thackray AC, Sammons RK, Macaskie LE. Yong P, Lugg H, Marquis PM. Bacterial synthesis of a calcium phosphate bon-substitute material. Journal of Materials Science Materials in Medicine. 2004;**15**,403-406. DOI: 10.1023/B:JMSM.0000021110.07796.6e

[84] Ledo HM, Thackray AC, Jones IP, Marquis PM. Macaskie LE, Sammons RL. Microstructure and composition of biosynthetically synthesised hydroxypatite. Journal of Materials Science Materials in Medicine. 2008;**19**:3419-3427. DOI: 10.1007/s10856-008-3485-3

[85] Terra J, Douado ER, Eon J-G, Ellis DE, Gonzalez G, Rossi AM. The structure of strontium-doped hydroxyapatite an experimental and theoretical study. Physical Chemistry Chemical Physics. 2009;**11**:568-577. DOI: 10.1039/b802841a

[86] Jensen J, Mackintosh AR. Rare earth magnetism. International Series Monographs in Physics. Birman J, Edwards SF, Llewellyn-Smyth CH, Rees M, editors. Oxford: Clarendon Press; 1991.http://www.fys.ku.dk/~jjensen/Book/Ebook.pdf [Accessed 2017-01-27]

[87] Schenk G, Ge Y, Carrington LE, Wynne CJ, Searle IR, Carroll BJ, Hamilton S, de Jersey J. Binuclear metal centers in plant purple acid phosphatises: Fe + Mn in sweet potato and Fe + Zn in soybean. Archives of Biochemistry and Biophysics. 1999;**370**:183-189. DOI: 10.1006/abbi.1999.1407

[88] Moriyama S, Mikheenko I, Macaskie LE. Magnetic Recovery of Neodymium in Bio-Hydroxyapatite. [Internet]. 2017.http://youtu.be/JIeC9nqeGpc [Accessed 2017-01-24]

[89] Singh, S. Use of *Serratia* sp. biofilm to recover rare earth element metals from model acidic waste solutions. [MSc Thesis] University of Birmingham UK; 2015.

[90] Murray AJ, Singh S, Vavlekas D, Tolley MR, Hathway TP, Boden R, Macaskie LE. A novel biomineralisation process for continuous selective recovery of rare earth elements. Proceedings of 53rd Annual Conference of Metallurgists, Vancouver Sept 28-Oct 1 2014 Paper 8669, 2014. Published by The Metallurgy & Materials Society of Canada, Quebec, Canada. ISBN: 978-1-926872-24-7

[91] Volesky B, editor. Biosorption of Heavy Metals. Boca Raton: CRC Press; 1990. p. 408

[92] Kotrba P, Mackova M, Macek T, editors. Microbial biosorption of metals. Dordrecht, Heidelberg, London, New York: Springer; 2011. p. 317. DOI: 10.1007/978-94-007-0443-5

[93] Paterson Beedle M, Readman JE, Hriljac JA, Macaskie LE. Biorecovery of uranium from aqueous solutions at the expense of phytic acid. Hydrometallurgy. 2010;**104**:524-528. DOI: 10.1016/j.hydromet.2010.01.019

[94] Mejare M, Bulow L. Metal-binding proteins and peptides in bioremediation and phytoremediation of heavy metals. Trends in Biotechnology. 2001;**19**:67-73. DOI: 10.1016/S0167-7799(00)01534-1

[95] Kuroda K, Ueda M. Engineering of microorganisms towards recovery of rare metal ions. Applied Microbiology and Biotechnology. 2010;**87**:53-60. DOI: 10.1007/s00253-010-2581-8

[96] Horiike T, Yamasjita M. A new fungal isolate, *Penidella* sp strain T9 accumulates the rare earth element dysprosium. Applied and Environmental Microbiology. 2015;**81**:3062-3086. DOI: 10.1128/AEM.00300-15

Chapter 2

Hydrometallurgical Recovery Process of Rare Earth Elements from Waste: Main Application of Acid Leaching with Devised τ-T Diagram

Namil Um

Additional information is available at the end of the chapter

http://dx.doi.org/10.5772/intechopen.68302

Abstract

Rare earth elements (REEs) are essential ingredients for developing modern industry as well as designing and developing high technology products used in our daily lives. However, China monopolizes the supply chain for rare metals, and there have been growing concerns around limited supply in other countries. As such, there is demand for intensive research into the recovery of REEs from a large amount of the wastes produced in various industry links, including radioactive residues, polishing powders, catalysts, magnetic materials, batteries, etc., because they can play an important role in the resource supply. Therefore, in this chapter, we introduced the main application of acid-leaching process as a hydrometallurgical method to obtain the REEs from waste containing rare earths; the effective application of acid-leaching process is discussed through the leaching behavior of rare earths in acidic solution and the synthesis method of target rare earth during acid leaching for purifying it from non–rare earth. The devised $\tau - T$ diagram application to the hydrometallurgical method for selective acid leaching is also discussed using the leaching kinetics of rare earth in acidic solution and might be helpful to further corroborate the future plans for technical scale with recovery effectiveness of REEs.

Keywords: rare earth element, hydrometallurgical recovery process, acid leaching, leaching kinetics, devised τ − T diagram

1. Introduction

Many industries have become highly dependent on products that cannot be made without using rare earth metals. Although the importance of REEs is increasing, the supply and the price are not stable [1, 2]. For that reason, many trading corporations and manufacturers globally have been aware of the importance of REEs for the industrial economy and have been planning their strategies to secure a stable supply. As such, there is demand for intensive investigation into the recovery of rare earths from various types of wastes containing rare earth elements (REEs) as they should serve as a new source and can play a leading role in securing resources [3–5].

Indeed, a large amount of wastes containing REEs have been produced in various industry links, including disused hydrogen storage batteries, magnetic materials, rare metal catalysts, polishing powders, hydrometallurgy residues, rare metal radioactive residues, etc. Therefore, many researchers are beginning to take an interest in the recovery of REEs by chemical and physical methods [6–8].

In general, the pyrometallurgical and hydrometallurgical processes have been used by many researchers and other professionals in the field of the recovery of material of interest from various types of mineral. The main disadvantages of pyrometallurgical process are the severe corrosion problems (generation of dust and gas) and their high energy requirements, whereas the hydrometallurgical process is more environmentally suitable and economical to treat the target materials on a small scale [9, 10].

The leaching process with acidic solution, which is one of the hydrometallurgical processes including leaching, solvent extraction, ion exchange and precipitation, has some drawbacks in terms of the separation and recovery of materials of interest from undesired species, because another method after the leaching process is needed to separate the materials of interest. Nevertheless, the leaching process is simple, and its cost is not expensive to implement for the treatment of the target materials. Another reason for this is discussed in more detail in Section3.

Therefore, in this chapter, we will introduce the main application of the leaching process among the hydrometallurgical processes, as methods to improve or to help the effective recovery of REEs from waste. In addition, the possibility of the selective leaching method with the devised $\tau - T$ diagram is also discussed using the characteristics of leaching kinetics of rare earth oxides (REOs) on acid leaching.

2. Hydrometallurgical process for recovery of rare earth elements (REEs) from waste

Hydrometallurgical operation is an essential part of extractive metallurgy and utilized in various metal refining plants throughout the world. This is known to be a flexible, highly selective and environmentally friendly method for the treatment of raw materials. The principal processes employed during hydrometallurgical treatment of the resources mainly include leaching, solvent extraction, ion exchange and precipitation, which varies depending on the material of interest to be recovered. The basic processes used for REEs recovery from the waste are similar to hydrometallurgical methods.

2.1. Leaching

Leaching REEs from the waste is an important part of rare earth processing using hydrometallurgical route. Physically beneficiated concentrates are leached in a suitable lixiviant directly or after heat treatment to dissolve the metallic elements. The known processes range from acid leaching with H_2SO_4, HCl, HNO_3 to leaching with NaCl or $(NH_4)_2SO_4$ of ion adsorbed clays. Complete understanding of these processes is essential for applying them to develop more feasible methods for the recovery of REEs from the waste. Therefore, this chapter focuses primarily on the main application of the acid-leaching process as a hydrometallurgical method needed in order to recovery REEs from waste containing rare earths.

2.2. Solvent extraction

Solvent extraction is an important technique that can usually be employed to separate and extract individual rare earths to get their mixed solutions and compounds from leached solutions after the leaching process using different cationic, anionic and solvating extractants viz. D2EHPA, Cyanex 272, PC 88A, Versatic 10, TBP, Aliquat 336, etc. For example, D2EHPA is the most widely studied extractant for rare earth separation from nitrate, sulfate, chloride and perchlorate solutions. Saponified PC 88A has been reported for their separation from chloride solutions while tributyl phosphate (TBP), a solvating extractant that extracts their nitrates from the aqueous solutions.

2.3. Ion exchange and precipitation

During the ion exchange process, different ion exchange cationic or anionic resins are employed depending on the constituent of the aqueous solution using batch or continuous mode in column to extract rare earths from leached solutions with low rare earth concentration. Cation exchange can be primarily used to obtain REEs. The affinity of the exchanged ions for the cation exchanger depends majorly on the charge, size and degree of hydration of the exchanged ions. However, the mechanism of the processes involved in anion exchangers was much more complex and could not be explained clearly. REEs show little tendency to form anionic complexes with simple inorganic ligands.

3. Why acid leaching?

REEs have strong affinity for oxygen, and thus, their resources are principally present in oxidic form as REOs. In addition, the crystal structure of REOs has its own peculiarities such as polishing ability, mechanical strength, wear resistance, etc. Thus, the rare earth as oxides is increasingly establishing themselves, with unique applications in numerous fields, such as the catalysts, ceramics, phosphores, glass and polishing, as shown in **Table 1** [11]. After being used for numerous applications in various fields, a large amount of waste containing REOs is generated. And even when the waste with nonoxides of REEs is generated, the initial step with a calcinations process for hydrometallurgical recovery process, which does not liberate the material interest from host waste, is available to oxidize the REEs; when the waste is treated using a high temperature of at least 500°C, the REEs in remaining residue will be oxidized to REOs. Assuming that all REEs exist as an oxide, the leaching behavior of REOs in an acid solution will be the key for effective recovery of REEs.

Application	RE technology	Required RE	Materials	Consumption (%)
Catalysts	Petroleum refining, catalytic converter, diesel additives, chemical processing, industrial pollution scrubber	La, Ce, Pr, Nd	**REOs**	19
Ceramics	Capacitors, sensors, colorants, scintillators, refractories	La, Ce, Pr, Nd, Y, Eu, Gd, Bu, Dy	**REOs**, $LaCl_3$: Ce, Eu: **$(Y, Gd)_2O_3$**, Pr, Ce: Gd_2O_2S	6
Magnets	Disc drives, MRI, power generation, microphones and speakers, magnetic refrigeration	Nd, Pr, Tb, Dy	Alloys, e.g., $SmCo_5$, $Nd_2Fe_{14}B$, $Tb_xDy_{1-x}Fe_2$ (x~0.3)	21
Metallurgical alloys	NiMH batteries, fuel cells, steel, lighter flints, superalloys, aluminum/ magnesium	La, Ce, Pr, Nd, Y	Intermetallics, e.g., Lm $(Ni_{3.6}Mn_{0.4}Al_{0.3})$, $CeCoLn_5$, $YBa_2Cu_3O_7$	18
Phosphores	Display phosphors (CRT, LPD, LCD), fluorescent lighting, medical imaging, lasers, fiber optics	Eu, Y, Tb, Nd, Er, Gd, Ce, Pr	**REOs**, dopants, e.g., Ce_2O_5S: Tb^{3+}, $SrGa_2S_4$: Eu^{2+}, Tb: YAG, Tb: (La,Ca) PO_4	7
Glass and polishing	Polishing compounds, decolorisers/ colorisers, UV resistant glass, X-ray imaging	Ce, La, Pr, Nd, Gd, Er, Ho	**REOs**	22
Others	Nuclear, pigments	Eu, Gd, Ce, Y, Sm, Er, Ce, Y	**Gd_2O_3**, **Er_2O_3**, EuB_6, CeH_2, YH_3, **Y_2O_3**, **CeO_2**	7

Table 1. Applications of RE metals (Kim and Osseo-Asare, 2012) [11].

In terms of the cost consumption, the main advantages of the acid leaching process include lower capital (e.g., no expensive equipment and cheaper materials of construction) and lower operating costs (e.g., lower energy requirement and less maintenance). In addition, it is possible to apply other methods during acid leaching for purification of the material of interest (the reason for this is discussed in more detail in Section 4.2), and, the possibility for the selective leaching of REEs, using the characteristic of leaching kinetics, can be confirmed (the reason for selective leaching of REEs is discussed in more detail in Sections 5.2 and 5.3).

4. Acid leaching process for recovery of rare earth

4.1. Leaching behavior of rare earth oxides (REOs) on acid leaching

Cerium oxide with in a tetravalent state is the most stable phase among the rare earth oxides, so there are many difficulties associated with dissolving it in acidic solutions. For example, the E-pH diagram for Ce-H_2O system shown in **Figure 1a** indicates that the pH required for dissolution to Ce^{4+} from cerium oxide is in the range of < -0.8 at atmospheric pressure [12].

Therefore, the cerium oxide can be dissolved in highly concentrated acid solutions and at elevated temperatures. For confirmation of cerium oxide dissolution in sulfuric acid solution, two chemical reactions are considered to be involved in the conversion of cerium oxide into

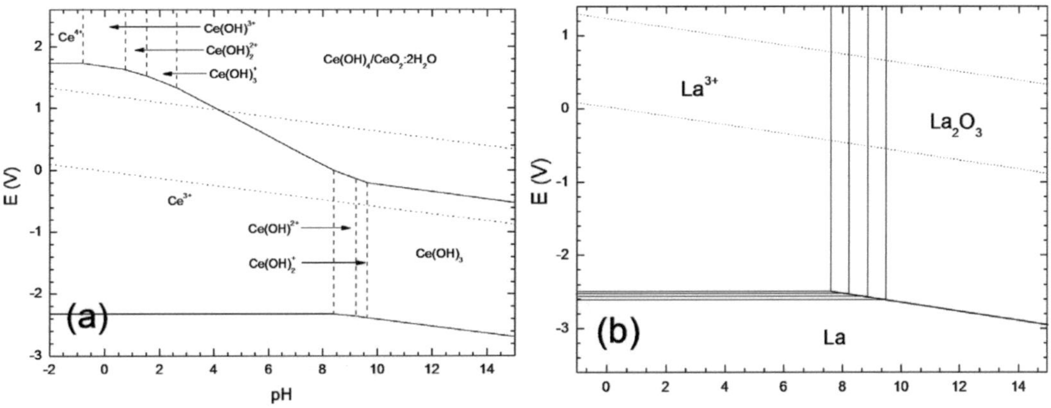

Figure 1. (a) An updated Poubaix diagram for cerium in aqueous perchlorate and (b) lanthanum in water (Scott et al., 2002) [12].

cerium sulfate in H_2SO_4-H_2O solution as follows. Cerium oxide reacts with sulfuric acid at atmospheric pressure according to the following dissolution reaction.

$$CeO_2 + 4H^+ \rightarrow Ce^{4+} + 2H_2O \quad (1)$$

At high sulfuric acid concentrations, the concentration of mixed cerium exceeds its solubility at an early stage. Then, cerium cations dissolved under saturated conditions directly form cerium sulfate as follows:

$$Ce^{4+} + 2SO_4{}^{2-} \rightarrow Ce(SO_4)_2 \quad (2)$$

Cerium (IV) cations can be reduced to cerium (III) cations as follows.

$$Ce^{4+} + 1/2\ H_2O \rightarrow Ce^{3+} + H^+ + 1/4\ O_2 \quad (3)$$

However, it can be estimated that this reaction proceeds much less than the cerium sulfate formation according to the following experimental results shown in **Figure 2a**. This result summarized by Um and Hirato (2012) [13] showed that it took more than 48hours to completely dissolve 0.02 mol CeO_2 powder with 2.5 μm average particle size in 100 ml of sulfuric acid (8 mol/dm^3) at 125°C. Comparing the XRD patterns of the precipitate (mixture of Ce-oxide and sulfate) provide evidence for the conversion of cerium oxide into cerium sulfate clearly (**Figure 2b**) and also support the results shown in **Figure 2a**; the intensity of the diffraction peaks originating from cerium sulfate increased with an increase in reaction time, while that from cerium oxide decreased.

Unlike cerium oxide, the dissolution of other REOs presenting as trivalent state occurred rapidly; for example, the E-pH diagram for lanthanum-water system shown in **Figure 1b** indicates that the pH required for dissolution to La^{3+} from lanthanum oxide is in the range of > at least 7.5 according to $La_2O_3 + 6H^+ \rightarrow La^{3+} + 3H_2O$. In addition, the results summarized by Um and Hirato (2013) [14] show that the dissolution of La, Pr, and Nd oxides took less than

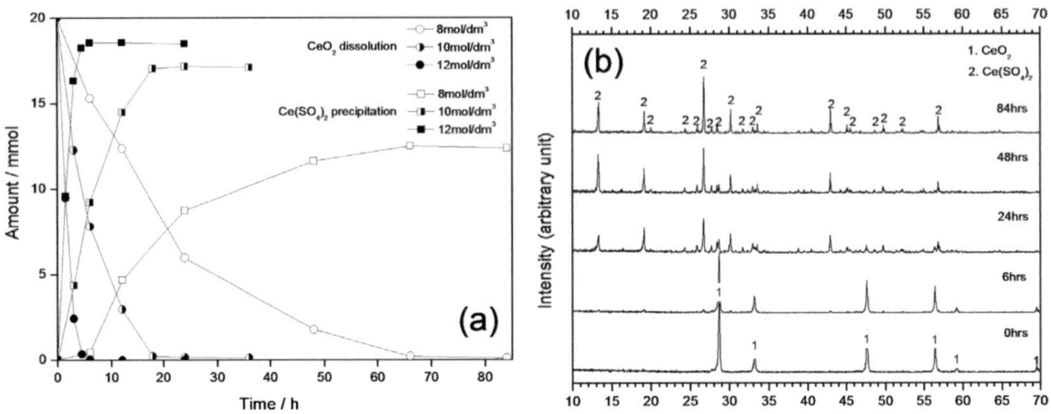

Figure 2. (a) Effect of sulfuric acid concentration on conversion kinetics of cerium oxide into cerium sulfate in acidic solution and (b) XRD patterns of the precipitate obtained at different reaction times (Um and Hirato, 2012) [13].

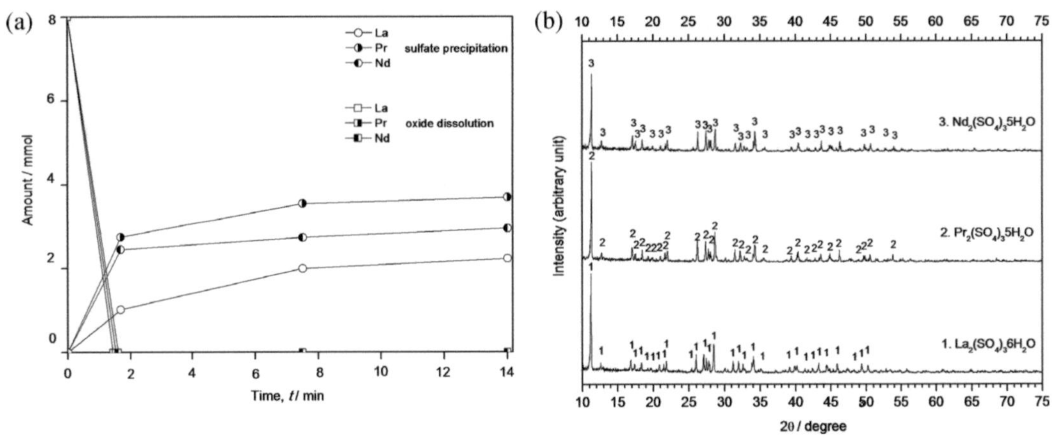

Figure 3. (a) Conversion kinetics of La_2O_3, Pr_2O_3, and Nd_2O_3 into La, Pr, and Nd sulfates in sulfuric acid solutions and (b) XRD patterns of La, Pr, and Nd sulfates obtained after 14-minute conversion kinetics (Um and Hirato, 2013) [14].

1 minute to dissolve in acid solution shown in **Figure 3a**. Under high initial amount of La_2O_3, Pr_2O_3, and Nd_2O_3 per sulfuric acid solutions, the concentration of dissolved La^{3+}, Nd^{3+}, and Pr^{3+} exceeds the solubility and these cations directly form sulfate. The results of the XRD patterns shown in **Figure 3b** provide evidence for that conversion. The chemical reactions below [Eqs. (4)–(9)] are considered to be involved in the conversion of La, Pr, and Nd oxides into precipitated sulfate in sulfuric acid solutions:

$$La_2O_3 + 6H^+ \rightarrow La^{3+} + 3H_2O \tag{4}$$

$$Pr_2O_3 + 6H^+ \rightarrow Pr^{3+} + 3H_2O \tag{5}$$

$$Nd_2O_3 + 6H^+ \rightarrow Nd^{3+} + 3H_2O \tag{6}$$

$$2La^{3+} + 3SO_4{}^{2-} + 6H_2O \rightarrow La_2(SO_4)_3 \cdot 6H_2O \tag{7}$$

$$2Pr^{3+} + 3SO_4{}^{2-} + 5H_2O \rightarrow Pr_2(SO_4)_3 \cdot 5H_2O \tag{8}$$

$$2Nd^{3+} + 3SO_4{}^{2-} + 5H_2O \rightarrow Nd_2(SO_4)_3 \cdot 5H_2O \tag{9}$$

4.2. Synthesis method for purification of rare earth during acid leaching

As mentioned in Section 1, the leaching process using an acidic solution for the recovery of rare earth from the waste has its demerits with respect to separation and recovery. The materials of interest are dissolved along with undesired species in the solution, and another method is thus needed to separate these materials; if waste containing REEs is dissolved in acid solutions, difficulties in purifying REEs from non–rare earth ions inevitably arise. Therefore, the synthesis method for purification of REEs during acid leaching is one of the simpler separation methods. According to Um and Hirato (2012 and 2016) [15, 16], for purification of target rare earth from rare earth polishing powder wastes, the REEs can be precipitated in the form of rare earths and sodium double sulfate ($NaRe(SO_4)_2{\cdot}xH_2O$) through the addition of Na_2SO_4 to H_2SO_4-H_2O solutions during the acid leaching because $NaRe(SO_4)_2{\cdot}xH_2O$ is poorly soluble under acidic conditions.

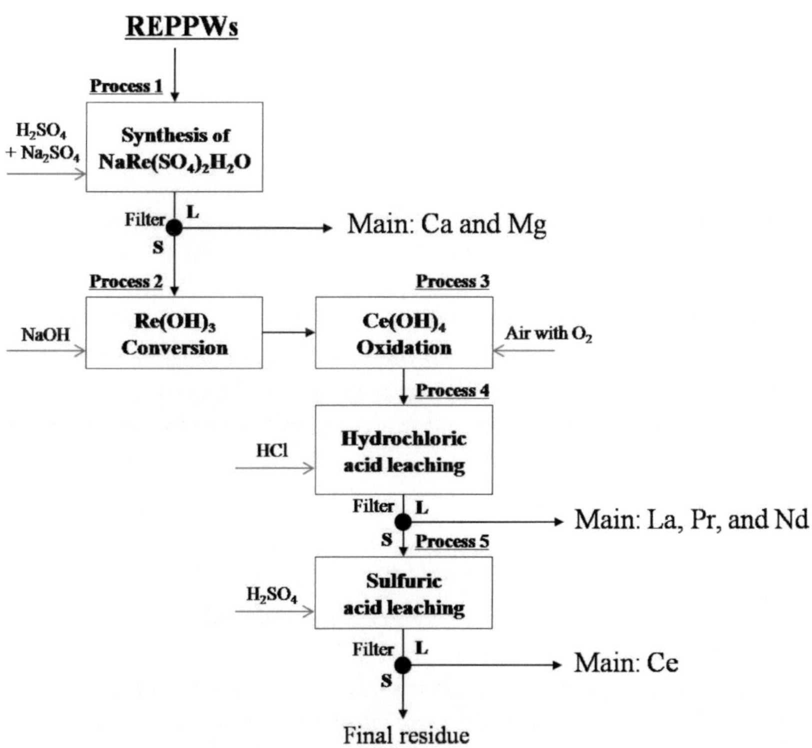

Figure 4. Schematic flow diagram for separation of rare earth elements from rare earth polishing powder wastes via a devised hydrometallurgical process (Um and Hirato, 2016) [16].

A five-process strategy for separation of REEs from polishing powder wastes in **Figure 4** was employed in the result from Um and Hitato (2016) [16]: process 1, the high yield of synthesized $NaRe(SO_4)_2{\cdot}xH_2O$ from REOs in Na_2SO_4-H_2SO_4-H_2O would be controlled by the function of the reaction temperature, sulfuric acid concentration, and Na_2SO_4 concentration; process 2, $NaRe(SO_4)_2{\cdot}H_2O$ is converted into $Re(OH)_3$ using sodium hydroxide (NaOH) solution; process 3, the oxidation of $Ce(OH)_3$ to $Ce(OH)_4$ via the injection of air including O_2 would be advantageous for the separation of cerium, existing as the main phase in polishing powder wastes, from other rare earths; process 4, the acid leaching with hydrogen chloride (HCl) at 25°C should be limited until the pH value is between 2.5 and 3.5 because a high yield of rare earths, except cerium, is obtained after acid leaching; process 5, the residue after process 4 is added to acid solution with H_2SO_4 to separate cerium from the final impurities. As shown in **Figure 4**, this result suggests the possibility of target lanthanide element from the other REEs using selective synthesis method. The separation of cerium from the other REEs can be carried out by selective synthesis of the tetravalent Ce-hydroxide in process 3, because the trivalent ceric ions, which is the major phase in polishing powder wastes, is the most likely to be oxidize to its tetravalent state by bubbling air with oxygen, unlike the other REEs presenting as trivalent cations.

5. Leaching kinetics and application of devised $\tau - T$ diagram

5.1. Leaching kinetics

The experimental data of REOs dissolution, which was performed under various conditions of acid concentration, reaction temperature, solid-to-liquid ratio, particle size, etc., were made to fit the shrinking core model to REO-dissolution vs. time curves, as follows.

The dissolution mechanism of B (solid) into A (acid-liquid) is described as a two-step process; the first step is B dissolution into A, and the second step is that the B cations dissolved under saturated conditions directly form B-precipitate in an aqueous acid medium, as mentioned in Section 4.1. Because these steps occur consecutively, if one is slower than the other, the step becomes rate determining. The formation of B-precipitate under saturated conditions is dependent on the acid concentration, the reaction temperature, etc., and its rate is much faster than the dissolution rate. Therefore, assuming that this reaction does not affect the dissolution rate, the rate of the leaching reaction between the B solid and the A liquid becomes the rate-determining step:

$$\text{A (acid} - \text{liquid)} + \text{bB (solid)} \rightarrow \text{Product of B cations in liquid} \tag{10}$$

According to Wadsworth and Miller (1979) [17], Liquid-solid reaction kinetics of dense (nonporous) particles is described by the most widespread shrinking core model. In the liquid-solid heterogeneous system, the reaction rate may be controlled as certain individual steps such as diffusion through the fluid film, product layer diffusion control, and surface chemical reaction. The experimental data of REOs dissolution, which was performed under various conditions of reaction temperature, acid concentration of the liquid, solid-to-liquid ratio, and particle size, were made to fit the shrinking core model with surface chemical

http://dx.doi.org/10.5772/intechopen.68302

reaction to REOs-dissolution vs. time curves according to Um and Hirato (2012 and 2013) [13, 14]. This result shows evidence of the leaching kinetics of REOs to fit the shrinking core model, which is discussed in detail below.

If the leaching reaction between the B solid and the A acid-liquid becomes the rate-determining step so that the reaction is followed by the shrinking core model with surface chemical reaction as shown in **Figure 5**, the rate of disappearance of B ($-r_B$; mol/m^{2}h) can be described as follows:

$$-r_B = \frac{1}{S_t}\frac{dN_B}{dt} \tag{11}$$

S_t(m^2) denotes the surface area of the B; dN_B (mol), the amount of B disappearing; in particular, if $dN_B \rightarrow bdN_A$(disappearing acid concentration of A; mol) according to the chemical reaction and their stoichiometric coefficient in Eq. (10), it is possible to obtain the expression with k_1 and C_A:

$$-\frac{1}{4\pi r^2}\frac{dN_B}{dt} = -\frac{b}{4\pi r^2}\frac{dN_A}{dt} = bk_1C_A \tag{12}$$

Here, k_1(h^{-1}) is the rate constant of first-order reaction controlled by the interaction between the A and the surface of B, and C_A(mol/m^2) is the concentration of A existing on the surface of B. Also, the amount of B disappearing can be expressed using the density and the volume of B as follows:

$$dN_B = \rho dV_t = \rho \cdot d\left(\frac{4}{3}\pi r^3\right) = \rho 4\pi r^2 dr \tag{13}$$

ρ(mol/m^3) presents the density of B; V_t (m^3), the volume of B (solid) after dissolution time t; r (m), the radius size of Bafter dissolution time t. According to Eqs. (12) and (13), Eqs. (14) and (15) can be written as:

$$-\frac{1}{4\pi r^2}\rho 4\pi r^2\frac{dr}{dt} = -\rho\frac{dr}{dt} = bk_1C_A \tag{14}$$

and

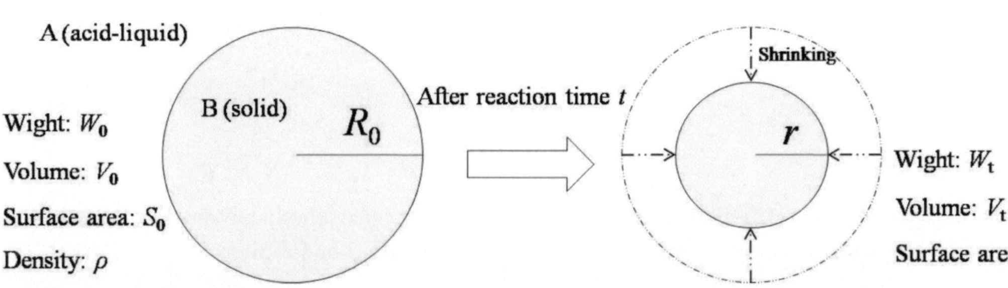

Figure 5. Shrinking core model with surface chemical reaction between A (acid-liquid) and B (solid).

$$\rho \int_{R}^{r} dr = bk_1 C_A \int_{0}^{t} dt \tag{15}$$

Integrating, yield the following expression for time t:

$$t = \frac{\rho}{bk_1 C_A}(R_0 - r) \tag{16}$$

If $r=0$, the time for complete dissolution reaction (τ; h) can be expressed as:

$$\tau = \frac{\rho R_0}{bk_1 C_A} \tag{17}$$

Combining these two Eqs. (16) and (17) in the form with the radius sizes of B before and after dissolution time ($\frac{r}{R_0}$) gives the expression for $\frac{t}{\tau}$ and k(apparent rate constant; h^{-1}):

$$\frac{t}{\tau} = 1 - \frac{r}{R_0} \tag{18}$$

and

$$k = \frac{1}{\tau} = \frac{bC_A}{\rho R_0} k_1 \tag{19}$$

For calculating k, it is necessary to confirm the change of the radius size of B before and after dissolution time. However, it is not easy to measure the size of target particle because of their heterogeneous spherical shape. Therefore, Eq. (20) can lead to advanced means of obtaining much easier measurement with the weight of B rather than that with the radius size.

$$\frac{W_t}{W_0} = \frac{\frac{W_t}{\rho}}{\frac{W_0}{\rho}} = \frac{V_t}{V_0} = \frac{\frac{4}{3}\pi r^3}{\frac{4}{3}\pi R_0^3} = \left(\frac{r}{R_0}\right)^3 \tag{20}$$

Here, W_0 and W_t represent the initial and residual amount of B vs. dissolution time t, respectively. X_t represents the dissolved fraction vs. dissolution time t, which was then calculated as follows:

$$X_t = 1 - \frac{W_t}{W_0} \tag{21}$$

Combining these Eqs. (18)–(21), the rate equation of surface chemical reaction controlled process for heterogeneous spherical particles of B is established as follows:

$$kt = 1 - (1 - X_t)^{\frac{1}{3}} \tag{22}$$

The rate constant k varies with experimental conditions and is the function of reaction temperature, acid concentration, particle size, and initial amount of B (target REEs) per A (acid-liquid) (B/A). When B is dissolved into acid-liquid in all experiments, with the dissolution rate constant, considered as the function of reaction temperature, sulfuric acid concentration, particle size, and B/A, can be expressed by the following equation:

$$k = k_0' e^{-\mathrm{Ea}/RT} C^m p^n S^l \tag{23}$$

where Ea is the activation energy (kJ/mol); R, the ideal gas constant, 8.314×10^{-3}(kJ/molK); T, the reaction temperature (K); C, sulfuric acid concentration (mol/dm^3); p, particle size (μm); S, B/A (mol/dm^3); k_0', pre-exponential factor; and m, n, and l are constants.

It can be seen that $1 - (1 - X_t)^{\frac{1}{3}}$ and time t are in a linear relationship and the slope of these line is k according to Eq. (22). In addition, the ln$k$ values calculated from these k values were plotted against $1/T$, lnC, ln$p$, and ln$S$ using equations $k = k_1' e^{-\mathrm{Ea}/RT}$ (relationship between rate constant and activation energy with reaction temperature), $k = k_2' C^m$ (rate constant and acid concentration constant), $k = k_3' p^n$ (rate constant and particle size constant), and $k = k_4' S^l$(rate constant and B/A constant) according to Eq. (23) where k_1', k_2', k_3', and k_4' are the pre-exponential factors. It can also be seen that ln$k$ and $1/T$ (or lnC, ln$p$, and ln$S$) are in a linear relationship and the slope of this line is $-\mathrm{Ea}/RT$(or m, n, and l).

The experimental data obtained from the study of Um (2012) [18], which shows the effect of reaction temperature, sulfuric acid concentration, initial amount of cerium oxide per sulfuric acid solutions (B/A), and particle size on the conversion kinetics of cerium oxide in sulfuric acid solutions, were fitted with theoretical functions derived from Eqs. (22) and (23). **Figure 6a** shows the effect of reaction temperature on the dissolution rate of cerium oxide. Such results indicate that $1 - (1 - X_t)^{\frac{1}{3}}$ and time t are in a linear relationship; the rate constants are calculated as slopes of the straight lines. The rate constants were used to determine the Arrhenius plot between ln$k$ and $1/T$ (application of $k = k_1' e^{-\mathrm{Ea}/RT}$) in **Figure 6b**. As shown in **Figure 6b**, the increase in dissolution rate constant with increasing temperature obeyed the Arrhenius equation with an activation energy of 123kJ/mol. The results for the effect of sulfuric acid concentration were applied to the kinetic model and rate constants for various sulfuric acid concentrations (8, 10, and 12 mol/dm^3) were determined by using the linear regressions in **Figure 6c**. A plot of ln$k$ versus lnC in **Figure 6d** shows that the constant m was calculated to be 6.54. This large constant indicates a strong effect of acid concentration on the dissolution rate. The slops in **Figure 6e** determined the rate constant related to the particle sizes (2.5, 10.0, 47.5, and 112.5 μm). As seen in **Figure 6e**, the decrease in particle size increased the dissolution rate of cerium oxide. The plot of ln$k$ versus ln$p$ in **Figure 6f** shows that the constant n was calculated to be −0.78. In addition, the linear regressions in **Figure 6g** were used to calculate the reaction rate constants for various C/S (0.04, 0.12, and 0.2 mol/dm^3). And the plot of ln$k$ versus ln$S$ in **Figure 6h** shows that the constant l was calculated to be −0.03, indicating that C/S has no effect on the dissolution rate of cerium oxide. Therefore, the kinetics equation on the dissolution of cerium oxide in sulfuric acid solution was $k = 3.26 \times 10^8 e^{-14800/T} C^{6.54} p^{-0.78} S^{-0.03}$.

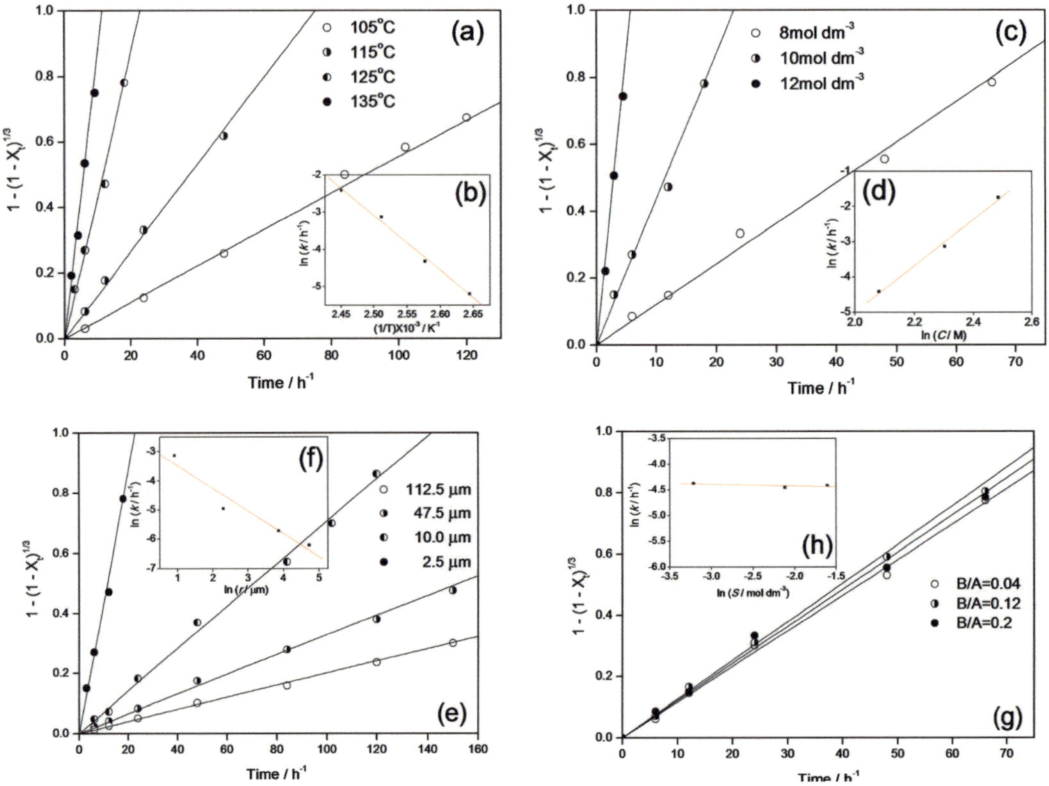

Figure 6. Plots of $1-(1-X_t)^{1/3}$ versus reaction time for (a) reaction temperatures, (c) different sulfuric acid concentrations, (e) particle sizes, and (g) B/A, and effect of (b) reaction temperature, (d) sulfuric acid concentration, (f) particle size, and (h) B/A on the conversion rate constants (Um, 2012) [18].

5.2. Devised $\tau - T$ diagram

The relationship between the time for complete dissolution reaction (τ) and the reaction temperature (T) will be the effective methods in terms of separation and recovery of REEs from the waste during the leaching process. In the case of a REOs-H_2SO_4-H_2O system, the relationship between τ and T can be constructed using the kinetics equation relating to the rate constant k with various functions of reaction temperature, sulfuric acid concentration, particle size, initial amount of REOs per acid solution, etc. If this relationship is dependent on the acid concentration, which affects the dissolution rate between the undesired species and the materials of interest (REOs), as well as the reaction temperature, the relationship can be devised using the equation of $k = k_5' e^{-\mathrm{Ea}/RT} C^m$ (k_5'; the pre-exponential factor) obtained from Eq. (23).

Rearranging the equation of $k = k_5' e^{-\mathrm{Ea}/RT} C^m$, one obtains:

$$\ln k = -\left(\frac{\mathrm{Ea}}{R}\right)\frac{1}{T} + (\ln k_5' + m\ln \mathrm{C}) \tag{24}$$

The equation shows that $\ln k$ and $1/T$ have a linear relationship, and the slope of this line is $-(\frac{\mathrm{Ea}}{\mathrm{R}})$. Putting the activation energies of the undesired species ($\mathrm{Ea}_{\mathrm{us}}$) and material of interest

(Ea_{int}) and the acid concentration constants of the undesired species (m_{us}) and material of interest (m_{int}) into the Eq. (24) respectively, the expressions between ln$k$ and $1/T$ can be obtained as follows:

$$\text{Undesired species}: \ \ln k = -\left(\frac{Ea_{us}}{R}\right)\frac{1}{T} + (\ln k'_5 + m_{us}\ln C) \tag{25}$$

and

$$\text{Material of interest}: \ \ln k = -\left(\frac{Ea_{int}}{R}\right)\frac{1}{T} + (\ln k'_5 + m_{int}\ln C) \tag{26}$$

Converting ln$k$ into the time for complete dissolution reaction ($\tau = e^{-\ln k}$) according to Eq. (19), devised $\tau - T$ diagrams can be obtained as follows:

$$\text{Undesired species}: \ \tau = e^{\left[\left(\frac{Ea_{us}}{R}\right)\frac{1}{T} - (\ln k'_5 + m_{us}\ln C)\right]} \tag{27}$$

and

$$\text{Material of interest}: \ \tau = e^{\left[\left(\frac{Ea_{int}}{R}\right)\frac{1}{T} + (\ln k'_5 + m_{int}\ln C)\right]} \tag{28}$$

Two types of $\tau - T$ diagram can be explained as shown in **Figure 7**.

In Type 1, the linear relation between ln$k$ and $1/T$ of undesired species [Eq. (25)] lies with that of material of interest [Eq. (26)] over the whole range of $1/T$; the ln$k$ values of Y-axis are converted into τ, whereas Type 2 shows that the exponential function related to τ and $1/T$ of Eqs. (27) and (28) is placed over the range considered. As shown in **Figure 7**, the increasing difference in τ between two linear relations (or exponential function curves) in the $\tau - T$ diagram can lead the material of interest to be separated from the undesired species by leaching.

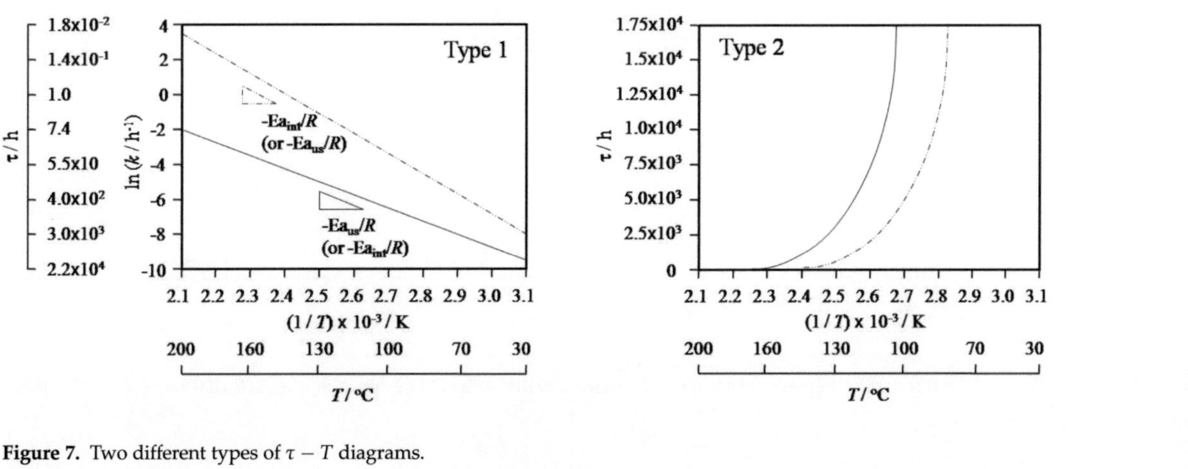

Figure 7. Two different types of $\tau - T$ diagrams.

From the point of view of the linear relation, the various cases of the relationship between the undesired species and the material of interest in the devised $\tau - T$ diagram of Type 1 can be expected as follows:

Case 1: $Ea_{int} \neq Ea_{us}$ and $m_{us} \doteq m_{int}$

Case 2: $Ea_{int} \neq Ea_{us}$ and $m_{int} \neq m_{us}$

Case 2-1: ($Ea_{int} > Ea_{us}$ and $m_{int} > m_{us}$) or ($Ea_{int} < Ea_{us}$ and $m_{int} < m_{us}$)

Case 2-2: ($Ea_{int} < Ea_{us}$ and $m_{int} > m_{us}$) or ($Ea_{int} > Ea_{us}$ and $m_{int} < m_{us}$)

Case 3: $Ea_{int} \doteq Ea_{us}$ and $m_{int} \neq m_{us}$

Case 4: $Ea_{int} \doteq Ea_{us}$ and $m_{us} \doteq m_{int}$

Because the difference in the activation energies as the slope of the linear relation in Eqs. (25) and (26) makes two lines angle inward toward the center, the difference in τ is definitely distinctive as the reaction temperature changes. For example, in Case 1, the difference in τ decreases with increasing reaction temperature, while, in Case 2-1, it increases. In Case 2-2, the wide difference between two materials can be observed over the whole range of temperature (T). If there is very little difference in the activation energies ($Ea_{int} \doteq Ea_{us}$), it is implied that the lines are nearly parallel with a given separation, as shown in Case 3 and Case 4; especially, in Case 4, the separation of the material of interest from the undesired species is difficult in terms of separation and recovery due to the similarity of the two lines (**Figure 8**).

The results from Um and Hirato (2013) [14] showing kinetics data of CeO_2 and Al_2O_3 dissolution performed under various conditions of sulfuric acid concentration and reaction temperature indicate that the $\tau - T$ diagram in CeO_2-Al_2O_3-H_2SO_4-H_2O system could be a good example of Case 3. As for the dissolutions of CeO_2 and Al_2O_3, the following equations were obtained as follows:

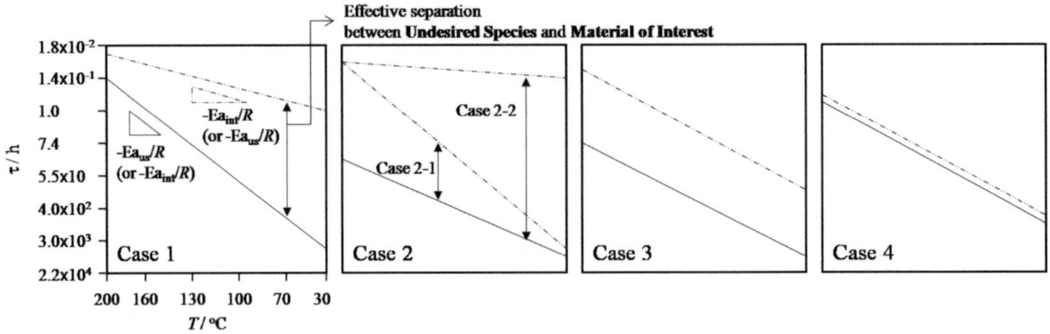

Figure 8. Various cases between the undesired species and the material of interest in the $\tau - T$ diagram (Um, 2012) [18].

CeO_2:

$$k = 1.43 \times 10^8 e^{-14800/T} C^{6.54} \tag{29}$$

Al_2O_3:

$$k = 1.23 \times 10^{13} e^{-15500/T} C^{0.41} \tag{30}$$

Putting the calculated $k_5' = 1.43 \times 10^8$ (CeO_2) and $k_5' = 1.23 \times 10^{13}$ (Al_2O_3), $\frac{Ea_{int}}{R} = 14800$ (CeO_2) and $\frac{Ea_{us}}{R} = 15500$ (Al_2O_3), and $m_{int} = 6.54$ (CeO_2) and $m_{us} = 0.41$ (Al_2O_3) into Eqs. (25) and (26), the expressions between ln$k$ and $1/T$ at various acid concentrations (C) can be obtained as follows.

$(1/T) \times 10^3$ from 2.1 to 3.1, here C is 8:

CeO_2:

$$\ln k = -14800/T + 32.4 \tag{31}$$

Al_2O_3:

$$\ln k = -15500/T + 31.0 \tag{32}$$

Here C is 10:

CeO_2:

$$\ln k = -14800/T + 33.8 \tag{33}$$

Al_2O_3:

$$\ln k = -15500/T + 31.1 \tag{34}$$

Here C is 12:

CeO_2:

$$\ln k = -\frac{14800}{T} + 35.0 \tag{35}$$

Al_2O_3:

$$\ln k = -15500/T + 31.2 \tag{36}$$

Here C is 14:

CeO_2:

$$\ln k = -14800/T + 36.0 \tag{37}$$

Al_2O_3:

$$\ln k = -15500/T + 31.2 \tag{38}$$

The revised $\tau - T$ diagram for a CeO_2-Al_2O_3-H_2SO_4-H_2O system can then be constructed as shown in **Figure 9** using the expressions given previously; the boiling points of H_2SO_4-H_2O solutions at different acid concentrations of 8, 10, 12, and 14 mol/dm^3 were calculated using OLI® software. Like Case 3, the two lines of CeO_2 (material of interest) and Al_2O_3 (undesired species) are nearly parallel with a given separation at each acid concentration. Particularly, the difference in τ increases with increasing acid concentration because cerium has a higher sensitivity of acid concentration than Al_2O_3; the large acid concentration constant of 6.54 indicates a strong effect of acid concentration on the dissolution rate of CeO_2, whereas Al_2O_3 (0.41) has less effect on the dissolution rate owing to changing acid concentration.

5.3. Devised $\tau - T$ diagram application to a hydrometallurgical method with acid leaching process

As mentioned in Section5.2, the devised $\tau - T$ diagram can be used to predict the dissolution behavior between material of interest and undesired species in the leaching process. As an example, Um and Hirato (2013) [14] reporting the recovery of cerium from rare earth polishing powder wastes presented the devised $\tau - T$ diagram application to a hydrometallurgical method with acid leaching process.

As shown in **Figure 10**, this method involved a two-stage leaching process in series with sulfuric acid solution, which was efficient to dissolve the main elements selectively and then

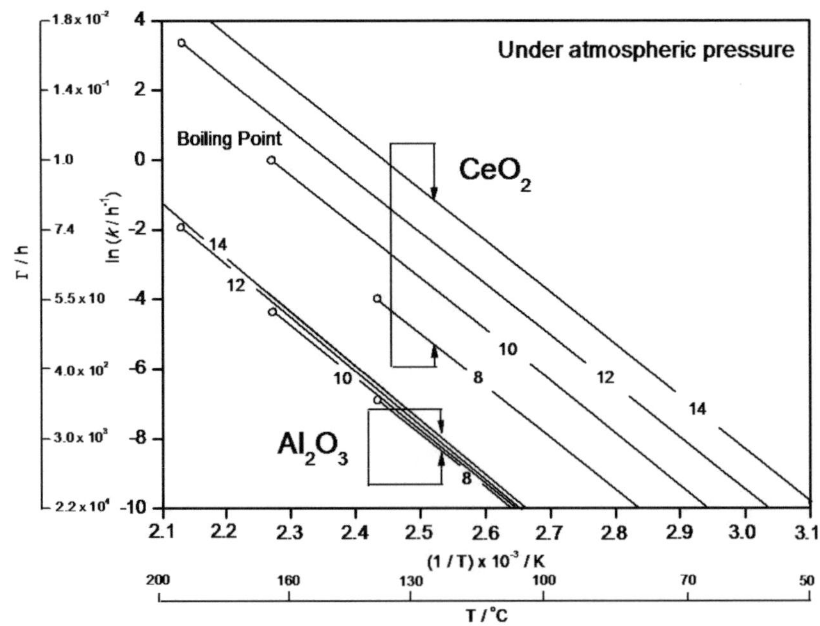

Figure 9. Revised $\tau - T$ diagram for CeO_2-Al_2O_3-H_2SO_4-H_2O system (Um and Hirato 2013) [14].

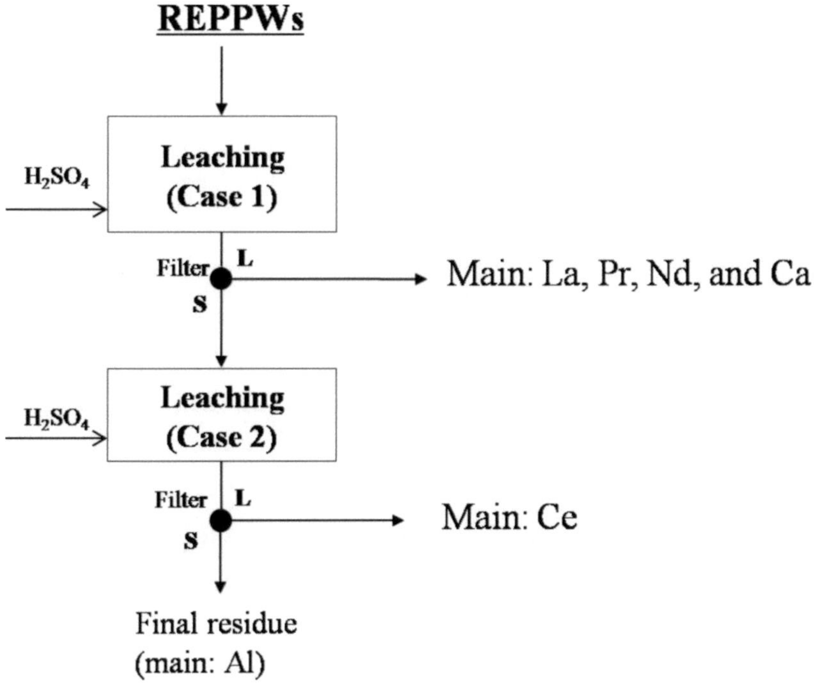

Figure 10. Schematic flow diagram for hydrometallurgical treatment of rare earth polishing powder waste for recovery cerium (Um and Hirato 2013) [14].

to recover cerium. Each dissolution experiment of stage 1 with La_2O_3, Pr_2O_3, Nd_2O_3, and CaO and stage 2 with CeO_2 with CeO_2 and Al_2O_3 by means of two-stage leaching was studied. In particular, after Step 1, CeO_2 can be dissolved leaving Al_2O_3 in a stable solid phase under dissolution conditions calculated using the revised $\tau - T$ diagram for a CeO_2-Al_2O_3-H_2SO_4-H_2O system shown in **Figure 9**; the increasing difference in τ between cerium and aluminum in the diagram leads CeO_2 to be separated from Al_2O_3 almost completely by leaching.

6. Future plan for technical scale

The perspectives of the studied hydrometallurgical recovery process of REEs from waste for potential chemical engineering applications could be discussed, including the preliminary estimation of energy consumptions and cost. For future plan of technical scale, the various conditions in lab-scale experiment, discussed in each section, including waste composition, reaction temperature, reaction time, chemical reagent, equipment etc, need to be considered as the factor of the preliminary estimation of energy consumptions and cost of the hydrometallurgical process in this study. For example, the reaction temperature and the acid concentration in synthesis of $NaCe(SO_4)_2 \cdot H_2O$ from CeO_2 in Na_2SO_4-H_2SO_4-H_2O solutions discussed in Section 4.2 and leaching of CeO_2 from the mixture of CeO_2 and Al_2O_3 in CeO_2-Al_2O_3-H_2SO_4-H_2O solution discussed in Section5.3 belong to a main factor of energy consumption:

1. On the basis of the temperature effect on the hydrometallurgical process with leaching and synthesis processes connecting with the reaction time, an appropriate control condition between temperature and reaction time on recovery method of REEs from waste will lead to the saving of energy consumption.

2. The control of acid concentration on the leaching process connecting with the reaction time and the precipitate after leaching (mentioned in Section 4.1) efforts toward the simplification of total process, and it will lead to savings in cost consumption.

3. The *in situ* recycling of acid solution, reaction reagent, etc. using in the hydrometallurgical process will be a cost-saving method.

4. Cheap and commercial available polyvinyl of polyvinyl chloride and polypropylene is an excellent material for recovery equipments of REEs, especially both reactor and filter in an acidic medium.

However, one concern is the energy- and cost-effectiveness of the hydrometallurgical recovery process of REEs from waste because, as mentioned in Section1, this depending on the rare earth price in world market being unstable. For example, China monopolizes the supply chain for rare earth, and there have been growing concerns over limited supply in other countries; many countries' trading companies and manufactures have been taking the initiative to secure other sources of rare metals. As such, we have to recognize the strategy to secure a stable supply of rare metals. Indeed, the race to obtain rare metals has started. Therefore, it is necessary to find more affordable alternatives to ensure the sustainability of the energy- and cost-effectiveness in the future plan for technical scale.

7. Conclusion

Hydrometallurgical processes, which mainly include leaching, solvent extraction, ion exchange and precipitation, are widely applied in many fields ranging from environmentally friendly methods for the treatment of raw materials to recovery of material of interest from waste. In the recovery field, the prominent application of leaching process among the hydrometallurgical processes can be the method to improve or to help the effective recovery.

In this chapter, we introduced the main application of acid leaching as a hydrometallurgical method needed in order to recovery rare earth elements (REEs) from waste containing rare earths because the leaching behavior of rare earth on acid solution can be the key to effective recovery as well as advantages including lower capital and lower operating costs. For purification of REEs, the synthesis method of target rare earth during acid leaching was also discussed; for example, the synthesis of rare earths and sodium double sulfate ($NaCe(SO_4)_2 \cdot H_2O$) is one of the simpler purification methods

In addition, the devised $\tau - T$ diagram application to the hydrometallurgical method for selective acid leaching was discussed. The devised $\tau - T$ diagram can be used to predict the dissolution behavior between REEs and undesired species in the leaching process using the results of leaching kinetics of rare earths in acid solutions. Therefore, this application can lead to advanced means of obtaining more effective recovery of REEs.

Author details

Namil Um

Address all correspondence to: inhaking96@hotmail.com

National Institute of Environmental Research, Incheon, Korea

References

[1] Wubbeke J. Rare earth elements in China: Policies and narratives of reinventing an industry. Resources Policy. 2013;**38**:384–394. DOI: 10.1016/j.resourpol.2013.05.005

[2] Baldi L, Peri M, Vandone D. Clean energy industries and rare earth materials: Economic and financial issues. Energy Policy. 2014;**66**:53–61. DOI: 10.1016/j.enpol.2013.10.067

[3] Tao X, Huiqing P. Formation cause, composition analysis and comprehensive utilization of rare earth solid wastes. Journal of Rare Earths. 2009;**27**:1096–1102. DOI: 10.1016/S1002-0721(08)60394-4

[4] Folgueras MB, Alonso M, Femandez FJ. Coal and sewage sludge ashes as sources of rare earth elements. Fuel. 2017;**192**:128–139. DOI: 10.1016/j.fuel.2016.12.019

[5] Funari V, Bokhari SNHB, Meisel T, Braga R. The rare earth elements in municipal solid waste incinerators ash and promising tools for their prospecting. Journal of Hazardous Materials. 2016;**301**:471–479. DOI: 10.1016/j.jhazmat.2015.09.015

[6] Reed DW, Fujita Y, Daubaras DL, Jiao Y, Thompson VS. Bioleaching of rare earth elements from waste phosphors and cracking catalysts. Hydrometallurgy. 2016;**166**:34–40. DOI: 10.1016/j.hydromet.2016.08.006

[7] Tunsu C, Petranikova M, Ekberg C, Retegan T. A hydrometallurgical process for the recovery of rare earth elements from fluorescent lamp waste fractions. Separation and Purification Technology. 2016;**161**:172–186. DOI: 10.1016/j.seppur.2016.01.048

[8] Yang X, Zhang J, Fang X. Rare earth element recycling from waste nickel-metal hydride batteries. Journal of Hazardous Materials. 2014;**279**:384–388. DOI: 10.1016/j.hazmat.2014.07.027

[9] Doyle FM. Teaching and learning environmental hydrometallurgy. Hydrometallurgy. 2005;**79**:1–14. DOI: 10.1016/j.hydromet.2004.10.022

[10] Jha MK, Kumari A, Panda R, Kumar JR, Yoo K, Lee JY. Review on hydrometallurgical recovery of rare earth metals. Hydrometallurgy. 2016;**165**:2–26. DOI: 10.1016/j.hydromet.2016.01.035

[11] Kim E, Osseo-Asare K. Aqueous stability of thorium and rare earth metals in monazite hydrometallurgy: Eh-pH diagrams for the systems Th-, Ce-, La-, Nd- (PO_4)-(SO_4)-H_2O at 25°C. Hydrometallurgy. 2012;**113–114**:67–78. DOI: 10.1016/j.hydromet.2011.12.007

[12] Scott A, Hayes PY, Thomas JO, Matthew JO, James OS. The phase stability of cerium species in aqueous systems: I. E-pH diagram for the Ce-$HClO_4$-H_2O system. Journal of the Electrochemical Society. 2002;**149**:C623–C630. DOI: 10.1149/1.1516775

[13] Um N, Hirato T. Precipitation of cerium sulphate converted from cerium oxide in sulphuric acid solutions and the conversion kinetics. Materials Transactions. 2012;**11**:1986–1991. DOI: 10.2320/matertrans.M-M2012826

[14] Um N, Hirato T. Dissolution behavior of La_2O_3, Pr_2O_3, Nd_2O_3, CaO and Al_2O_3 in sulfuric acid solutions and study of cerium recovery from rare earth polishing powder waste via two-stage sulfuric acid leaching. Materials Transactions. 2013;**54**:713–719. DOI: 10.2320/matertrans.M-M2013802

[15] Um N, Hirato T. Conversion kinetics of cerium oxide into sodium cerium sulphate in Na_2SO_4-H_2SO_4-H_2O solutions. Materials Transactions. 2012;**11**:1992–1996. DOI: 10.2320/matertrans.M-M2012827

[16] Um N, Hirato T. A hydrometallurgical method of energy saving type for separation of rare earth elements from rare earth polishing powder wastes with middle fraction of ceria. Journal of Rare Earths. 2016;**34**:536–542. DOI: 10.1016/S1002-0721(16)60059-5

[17] Wadsworth ME, Miller JD. Hydrometallurgical processes. In: Sohn HY, Wadsworth ME, editors. Rate Processes of Extractive Metallurgy. New York: Springer; 1979. pp. 133–244. DOI: 10.1007/978-1-4684-9117-3

[18] Um N. Development of recovery process of rare metals from various resources including rare metal waste by hydrometallurgical process [thesis]. Kyoto: Kyoto University; 2012

Section 2

Rare Earths Spectroscopy

Chapter 3

Optimisation of Parameters for Spectroscopic Analysis of Rare Earth Elements in Sediment Samples

Martin Makombe, Charlton van der Horst,
Bongiwe Silwana, Emmanuel Iwuoha and
Vernon Somerset

Additional information is available at the end of the chapter

http://dx.doi.org/10.5772/intechopen.68280

Abstract

The rapid demand for rare earth elements (REEs) in recent years due to increased use in various technological applications, agriculture, etc. has led to increased pollution and prevalence of REEs in the environment. Therefore, monitoring for REEs in the aquatic environment has become essential including the risk assessment to aquatic organisms. Since direct determination of REEs in sediment samples prove difficult at times, due to low concentrations available and complex matric effects, separation and enrichment steps are sometimes used. In this work, various REEs were determined employing wet acid digestion and lithium metaborate fusion in our optimised analytical technique. A comparison of the two analytical techniques was also made. The results obtained from the optimised ICP-OES radial view technique were in 5% agreement with the ICP-MS results from the same samples. The accuracy of the method was checked with the geological reference material GRE-03 and found to be in reasonable agreement. We demonstrated that there is a consistent relationship between the signals of the REEs and nebuliser gas flow rates, plasma power and pump speed. The detection limits for all the REEs ranged from 0.06 mg L^{-1} Yb to 2.5 mg L^{-1} Sm using the ICP-OES fusion technique.

Keywords: rare earths elements, spectroscopy optimization, lithium metaborate, fusion digestion and sediment

1. Introduction

Sediment forms the deposit of organic matter, silt, alluvium or sludge that settles at the bottom of a liquid. Water in rivers, wind and glaciers normally transports the sediment. The sediment forms the habitat of the aquatic ecosystem. Furthermore, it acts as a sink or source for metals that can contribute to biogeochemical processes that occur in the aquatic environment [1].

The nutrients may be either useful or harmful depending on the concentration. High concentrations of non-essential metals including rare earth elements (REEs) are toxic to the environment. In environmental impact studies, sediments play a crucial role in monitoring contamination levels. The elemental analysis of sediments and soils has become important in the environmental studies, particularly for the identification of the contaminants present in the matrix and the relative threshold levels of toxicity [2]. River and estuarine sediments can be used to assess the pollution level of heavy metals and REEs as the surface sediments interact with the water column and record the depositional pollution history.

A major issue in the determination of soil elemental constituents is represented by the refractory nature of many metals including the REEs. Traditional methods for the analysis of REEs in sediments, based on instrumental techniques, require the complete dissolution of samples. Typically, this is done through alkaline fusion, microwave digestion and acid digestion on hot plates at atmospheric pressure. Following dissolution of samples, the spectrometric measurement is performed using different spectrometric techniques, such as flame atomic absorption spectroscopy (FAAS), graphite furnace AAS (GFAAS), neutron activation energy (NAE), X-ray fluorescence (XRF), inductively coupled plasma optical emission spectrometry (ICP-OES), or ICP mass spectrometry (ICP-MS). Although the procedures of acid solubilisation are effective, they are time-consuming and can often result in the loss of the most volatile species [3, 4].

A prerequisite for the determination of rare earth metals and other refractory elements, such as chromium, zirconium, barium, titanium, hafnium and strontium, is a complete sample digestion [5, 6]. This is a huge problem in many analytical laboratories as the process consumes time and sometimes requires costly equipment. The digestion of rare earth and refractory elements is commonly achieved using hydrofluoric acid (HF) in combination with mineral acids, such as nitric acid (HNO_3), hydrochloric acid (HCl) and perchloric acid ($HClO_4$) [7]. The HF is good on breaking up the silica matrix and to liberate all entrapped metals. The acid digestion procedures also often result in incomplete analysis of refractory and rare earth metals. In addition, the use of HF is extremely dangerous due to the reactive nature of the acid [8]. Li et al. [9] reported the digestion of sediment samples by microwave technique using HNO_3-H_2O_2-HF reaching detection levels of ng/L on ICP-MS. An alternative technique to the wet acid digestion procedure is the fusion method. This is achieved by mixing the sample with flux such as lithium, sodium and potassium borate fluxes. Occasionally $NaCO_3$, K_2CO_3, Na_2O_2 and NaOH fluxes may be used [10, 11]. The use of the flux-fusion approach is preferred due to the following reasons: (i) it is a complete digestion technique; (ii) it is much safer because HF is not used; (iii) the dominance of flux in solution results in similar matrix composition perfectly homogenous; (iv) it maintains constant grain size, thus the sample obtained will be almost perfectly homogenous; (v) the solutions are stable in dilute nitric acid and (vi) sample preparation time is shorter than that of the conventional acid digestions [10, 11].

The use of lithium metaborate ($LiBO_2$) as flux is similar to the XRF preparation of making fused beads, where total dissolution of sample is required. The key objective of this chapter is the determination of optimum instrument parameters and sample preparation conditions for ICP-OES analysis. Thus, this will facilitate very precise, accurate and rapid measurements of rare earth elements in sediment samples. Two different procedures used in this chapter for sediments decomposition were open-vessel wet digestion and lithium tetraborate fusion techniques [5, 6, 12].

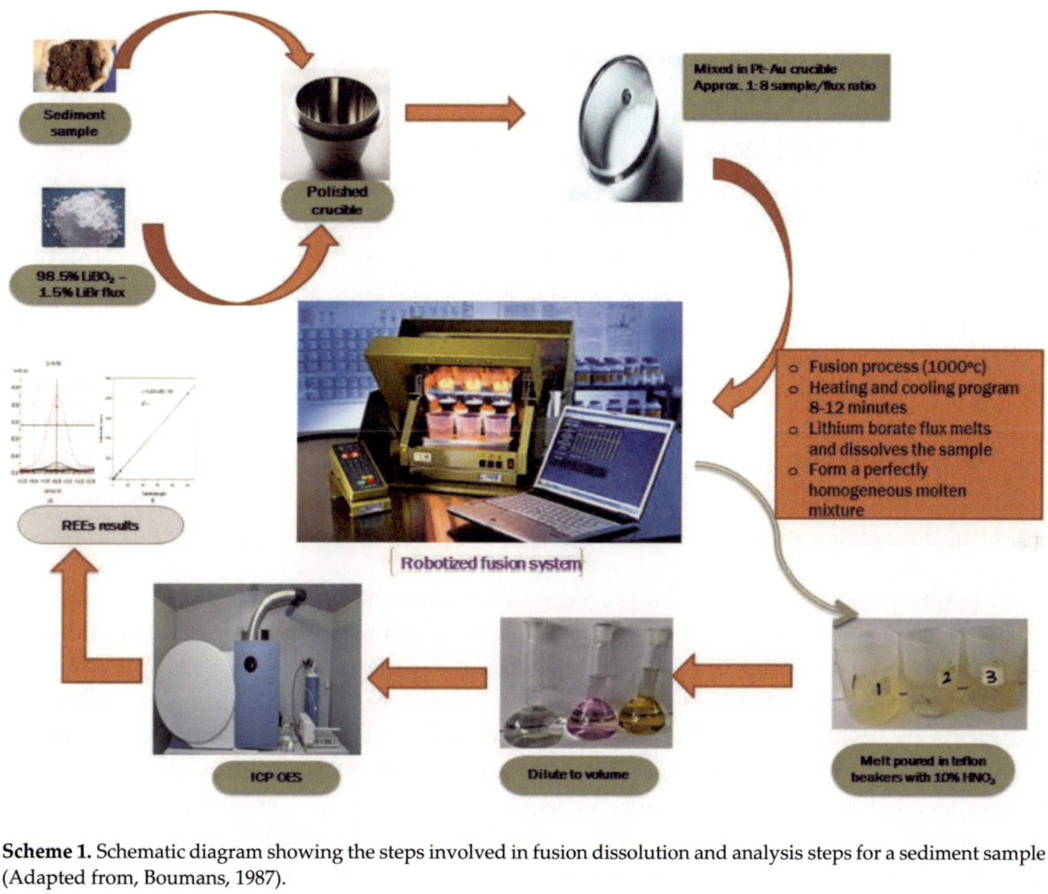

Scheme 1. Schematic diagram showing the steps involved in fusion dissolution and analysis steps for a sediment sample (Adapted from, Boumans, 1987).

2. Materials and methods

2.1. Sediment sample preparation by fusion for ICP analysis

Upon arriving at the laboratory, the samples are kept frozen to minimise the potential for volatisation or biodegradation between sampling and analysis. The preparation of sediment samples for chemical analysis collected from streams, rivers, ponds, lakes and oceans involves some steps to obtain a well-represented portion. Samples are first screened to remove foreign objects and large particulates, dried at 70°C and split to retain a fraction for storage. The chemically active fraction of sediment is usually cited as that portion which is smaller than 63 mm (silt + clay) fraction [13]. All weights were measured on a five-digit analytical balance (Mettler model).

A control sample can be prepared from portions that have been analysed. Thus, such finished material should be combined and well mixed into a larger storage container for future use as control sample that monitors the extraction process. Recovery test of the control and any reference material is evaluated after each run.

2.2. Wet digestion technique

For the wet digestion procedure, a mixture of HCl, HF HNO_3 and $HClO_3$ is added sequentially to the sample. All chemicals used for the preparation of standards and reagents were of analytical pure grade purchased from Merck (South Africa). The ideal sample weight for optimum dissolution is between 0.25 and 1.0 g weighed in Teflon beakers. For this procedure, HCl was boiled to near dryness. The sample was soaked with 1:1 HF and HNO_3 for 8 h or overnight at room temperature to enhance dissolution and then boiled to near dryness. A 1:1 mixture of HNO_3 and $HClO_3$ was boiled until all white dense fumes disappeared and this stage drove off all HF remaining. Another boiling of HNO_3 followed by 10% HNO_3 v/v to redissolve all salts completed the dissolution stage. After cooling, the sample was transferred to 50 ml volumetric flask and topped with 10% HNO_3 v/v. Mixing of the sample was achieved by vigorous shaking, and the clear solution was transferred to a glass vial for ICP-OES analysis [9, 14].

2.3. Lithium metaborate fusion technique

The lithium metaborate dissolution procedure comprises the powdered sample mixed with flux of high purity in a desired proportion. The sediments that have organic matter and volatile phases are first ignited to fully oxidise the matter. However, the loss on ignition (LOI) can result in some loss of alkali metals; therefore, it was avoided in this study and, again, the sediments analysed had minimum volatile matter. The accurately weighed sample and flux is heated to approximately 1000°C on an automated fusion fluxer machine. The temperature is controlled by the gas flow that can be increased or reduced. The optimum temperature will melt and dissolve the sample to a perfectly homogeneous mass. Upon completion of the dissolution stage, the molten mixture is poured into heat-resistant Teflon beakers containing 100 mL of 10% v/v HNO_3 or as per dilution of choice for ICP analysis. Alternatively, the molten solution is poured into a preheated mould to produce a glass disk for XRF analysis [11].

Basically, the standard fusion process consists of the following three steps: (i) melting the sample/flux mixture; (ii) pouring the molten glass into an acid solvent; and (iii) dissolution of the molten glass into solution [15, 16]. The three steps play a huge significance in optimisation of the fusion technique. If one or more of the steps are not properly adhered to, then incorrect results and poor precision is obtained. The sample to flux ratio is essential in making sure that the entire sample is thoroughly dissolved in solution but avoiding flux saturation.

2.4. Fusion instrumentation, borate-fused fluxes and platinum accessories

The complete fully automated M4 fusion instrument (Corporation Scientific Claisse Inc, Canada), a three-position automatic gas instrument, was used in this study. Different borate fluxes were tried and the lithium metaborate (98.5% $LiBO_2$–1.5% LiBr) was the best, due to higher solubility and faster crystallisation during cooling. The high-purity borate fluxes are fused and consist of homogenous, spherical and vitreous particles, which are water-free and non-hygroscopic. To avoid wetting of the crucibles, a pre-mixed flux with a non-wetting agent was employed. The halogenated compounds of iodine and bromine were used to make halo acids, for example, LiI, LiBr, NaBr, KI and so on, as they produce good releasing agents. Claisse's type crucibles used were of exceptional quality, made of 95% platinum alloyed with 5% gold. It is highly recommended to mirror-polish the crucibles to retain flat and smooth surfaces to aid easy transfer of the melt into the acid ICP mixture [11]. All the fluxes used and platinum ware were purchased from Claisse (Canada).

2.5. Dilution for ICP analysis

The dissolution procedure for flux-fusion preparation of sediments results in an aqueous solution that can be analysed in a single analytical session for REEs, major and trace elements. Most sediments contain trace amounts of REEs depending on the sampling site and level of pollution from anthropogenic sources. It is suggested that the minimum or optimum dilution be performed to get low detection limits and precise results be employed. The salt content in the solution requires a dilution that does not affect the mobility of the sample on the ICP instrumentation. Too viscous solution of the alkaline salt tends to deposit salts on the ICP torch and clogging of the nebuliser [17]. Different dilutions can be prepared, if the overall goal of generating enough analyte at an appropriate concentration is fulfilled. A 500 times (nominal) dilution of sample was considered safe in sediments analysis giving acceptable detection limits.

2.6. ICP-OES instrumentation

The Spectro Arcos ICP-OES (SPECTRO Analytical Instruments GmbH Boschstrasse 10, Germany) equipped with smart analyser software was initialised for about 20 min before analysis to get stable plasma. All standard solutions were prepared from high-purity 1000 mg/L ICP-grade standards. The Spectro Arcos ICP-OES is equipped with charge coupled device (CCD) detectors and side-on plasma interface (SPI) or commonly known as radial that provides high precision and stability for less sensitive requirements [18]. The sample injection mode was by continuous nebulisation, and the signal processing or line measurement was based on the peak height. Polynomial plotted mode corrected the background. The ICP-OES analyses were carried out in controlled room of 20 ± 2°C. The results were verified for accuracy by an independent laboratory using ICP-MS (Thermo-Fisher X-Series II quadrupole ICP-MS with a New Wave UP213 solid-state laser ablation system). The spot size of the laser can be set to various diameters between 10 and 300 microns.

With ICP-OES analysis, the multi-elemental analysis using the simultaneous or sequential optical systems with radial viewing of the plasma is possible. To generate plasma for excitation, the argon gas is supplied to the torch coil and high-frequency electric current is applied to the work coil at the tip of the torch tube. The high-frequency current generated by the electromagnetic field ionises the gas. When the excited atoms return to low energy position, emission rays (spectrum rays) are released, and the emission rays that correspond to the photon wavelength are measured. Samples are nebulised and the resulting aerosol is transported to the plasma torch in atomised state. Element-specific emission spectra are produced by radiofrequency (RF) inductively coupled plasma. Background correction is required for trace element determination. The ICP-OES measurement conditions need optimisation to get the best conditions for analysis, such as nebuliser flow rate, pumping speed, auxiliary and coolant flow rates and plasma power [19]. **Table 1** shows the optimised conditions for ICP-OES measurements equipped with a side-on plasma interface (SPI) or commonly known as radial [18]. ICP-OES and ICP-MS analyses are well regarded as appropriate environmental measurement techniques in measuring REEs, with the latter being more suitable for ultra-trace elemental levels [20]. Nowadays, due to the higher sensitivity achieved with axially viewed plasma and better spectral resolution given by high-resolution monochromators, it is expected that low concentrations of all naturally occurring lanthanides may be directly quantified by ICP-OES analysis [17].

Nebuliser type	Crossflow
Plasma power	1400 W
Torch	Demountable with alumina injector
Torch position	0
Nebuliser flow rate	0.8 L/min
Plasma flow rate	14.0 L/min
Auxiliary flow rate	2.1 L/min
Pump rate	40 rpm
Integration time	28 s
Replicates	3
Viewing height	Optimised on SBR
Plasma view orientation	Radial
PMT voltage	650 V
Background correction	Polynomial
Resolution	8.5 picometre
Detector	29 linear CCD

Table 1. Instrument operating conditions for the spectroscopic analysis of REEs.

2.7. Integration time

The integration (read) time was optimised using the instrument auto-integration mode. This phenomenon will take a snapshot of the intensities of various lines to be measured before the actual readings commence. The trace rare earth elements in sediments require larger integration time due to their low intensities and thereby increasing sample throughput. Additionally, longer read times lower the detection limits by reducing the effects of noise. Higher intensities will require shorter read time to reduce memory effects [9, 17].

2.8. Calibration

The calibration standards were prepared by diluting the stock multi-elemental standard solution 1000 mg L^{-1} in 10% (v/v) HNO_3 and 1.5% w/v flux (1.5 g in 100 mL). The calibration curves for all the studied elements were in the range of 0.01–1.0 mg L^{-1}. External calibration strategy is preferred for analysis of larger number of samples whilst standard addition method is suitable for small batches of samples. To avoid bias in external calibration method, a perfect matrix match of standards and samples is a prerequisite. Affected to a lesser degree by changes of the matrix composition or the presence of easily ionisable elements, calibration functions with excellent linearity and correlation were obtained, even without the use of an internal standard and an ionisation buffer [9, 17].

2.9. Interferences

The emission intensities were obtained for the most sensitive lines minimum of spectral interference. The REEs have very complex emission spectra and are difficult to measure in the presence of one another at very high concentrations using ICP-OES analysis. For this reason, ICP-MS analysis is the preferred measurement technique, but it can be very costly. Nebuliser, chemical, ionisation and spectral interferences are all present in ICP systems, but spectral interferences are most prominent [21].

Spectral interferences are common in ICP analysis and result from the overlapping profiles of spectral and interfering lines. The stray light from line emission of high concentration elements or particulate matter from atomisation process gives rise to enhanced emission light. This phenomenon gives rise to background emission, which is unwanted signal in analysis. This can be compensated for by subtracting the background emission determined by measuring the emission level on the two sides of the analyte peak and subtracting their average from the peak value [22].

Spectral interferences are caused by background emission from continuous or recombination phenomena, stray light from the line emission of high concentration elements, overlap of a spectral line from another element or unresolved overlap of molecular band spectra. Background emission and stray light can usually be compensated for by subtracting the background emission determined by measuring the emission level on the two sides of the analyte peak and subtracting their average from the peak value [22].

Figure 1 shows an example of the position of the atomic absorption peak and the background peak, as obtained for unidentified element. Spectral overlaps may also be avoided by using an alternate wavelength or can be compensated by equations that correct for inter-element contributions.

Nebuliser interferences, commonly known as matrix effects, arise from physical and chemical differences between analytical standards and samples. The inconsistent presence of matrix salts, different viscosities and surface tension of the liquid between the samples are

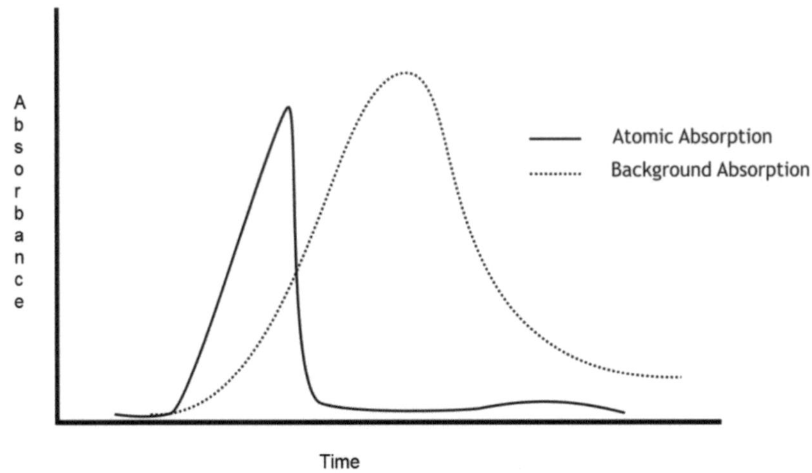

Figure 1. The position of the atomic and background absorption peaks that may cause spectral overlaps and contribute to inter-element interferences.

common problems of nebuliser interferences. The high salt content used in fusion technique needs to be diluted to avoid these interferences and match matrix of standards and samples. Using a high-solid nebuliser, the use of peristaltic pump can also reduce nebuliser interference [23].

2.10. Analytical wavelengths selection

Several criteria were applied for the selection: sensitivity (intensity of the line), selectivity (lack of interferences of other lines) and limits of detection (LOD). All selected lines were tested by using individual 0.1 mg L^{-1} REEs standard solutions in 10% HNO_3–$LiBO_2$ solution, searching the peak-centre in a first step. The optimised results obtained for ICP-OES analysis are shown in **Table 2**. Wavelength selection is somewhat of an individual choice that commonly varies from analyst to analyst and matrix of samples. However, there is a developing consensus regarding the wavelengths best suited for a target analyte [17, 21].

2.11. Detection limits

Method detection limits (MDL) for each element were calculated and are shown in **Table 2**. For comparison purposes, the LODs obtained using multi-acids and fusion dissolution for ICP-OES analyses are also shown. Validation was performed in terms of limits of detection (LOD), limits of quantification (LOQ), linearity, precision and recovery test. Lacking a suitable certified reference material, trueness was estimated using the recovery rate on fortified samples. Standards and blanks should be prepared using the same $LiBO_2$ strength and acids used for the samples and should be spiked with NIST traceable single and multi-element standards. The whole procedure should also be checked with certified reference material like the Geostart GRE-03 used in this study to ensure accuracy. The method detection

Element (ppm)	Line (nm) [ICP-OES]	MDL (mg L^{-1}) (Acid digestion)	MDL (mg L^{-1}) (Fusion digestion)
La	333.749	1.1	0.3
Ce	418.660	1.6	1.0
Pr	417.939	1.2	1.4
Nd	430.358	2.4	2.1
Sm	360.428	2.8	2.5
Eu	420.505	0.8	0.1
Gd	335.862	1.1	0.7
Tb	231.890	2.3	1.9
Dy	338.502	1.4	0.7
Ho	345.600	0.8	0.3
Er	349.910	0.5	0.2
Tm	313.126	0.5	0.3
Yb	369.419	0.1	0.06
Lu	261.542	0.1	0.04
Y	371.030	0.1	0.08

Table 2. Wavelengths used and detection limits obtained for the use of acid and fusion digestion methods in ICP-OES analysis.

limits (MDLs) were calculated based on relative standard deviation (RSD) of 10 consecutive measurements of the matrix blank. The signals to background ratio, background equivalent concentration (BEC) and LOD were calculated according to the three equations provided below [24–26]:

$$SBR = \frac{[\,Int(standard) - Int(spectral\,background)\,]}{Int(spectral\,background)} \tag{1}$$

$$BEC = \frac{C(standard)}{SBR} \tag{2}$$

$$C_{DL} = 3 \times RSD_b \times \frac{BEC}{100} \tag{3}$$

where

RSD_b is the relative standard deviation of spectral background intensity (10 replicates of blanks); C is the concentration of the standard; SBR is the signal to background ratio and BEC is the background equivalent concentration.

3. Results and discussion

3.1. Sample-flux ratio optimisation

The influence of the flux/sample ratio on the concentration of the REEs is illustrated in **Figure 2**. The weight of the sample was examined from 0.100 to 0.300 g. The flux was varied

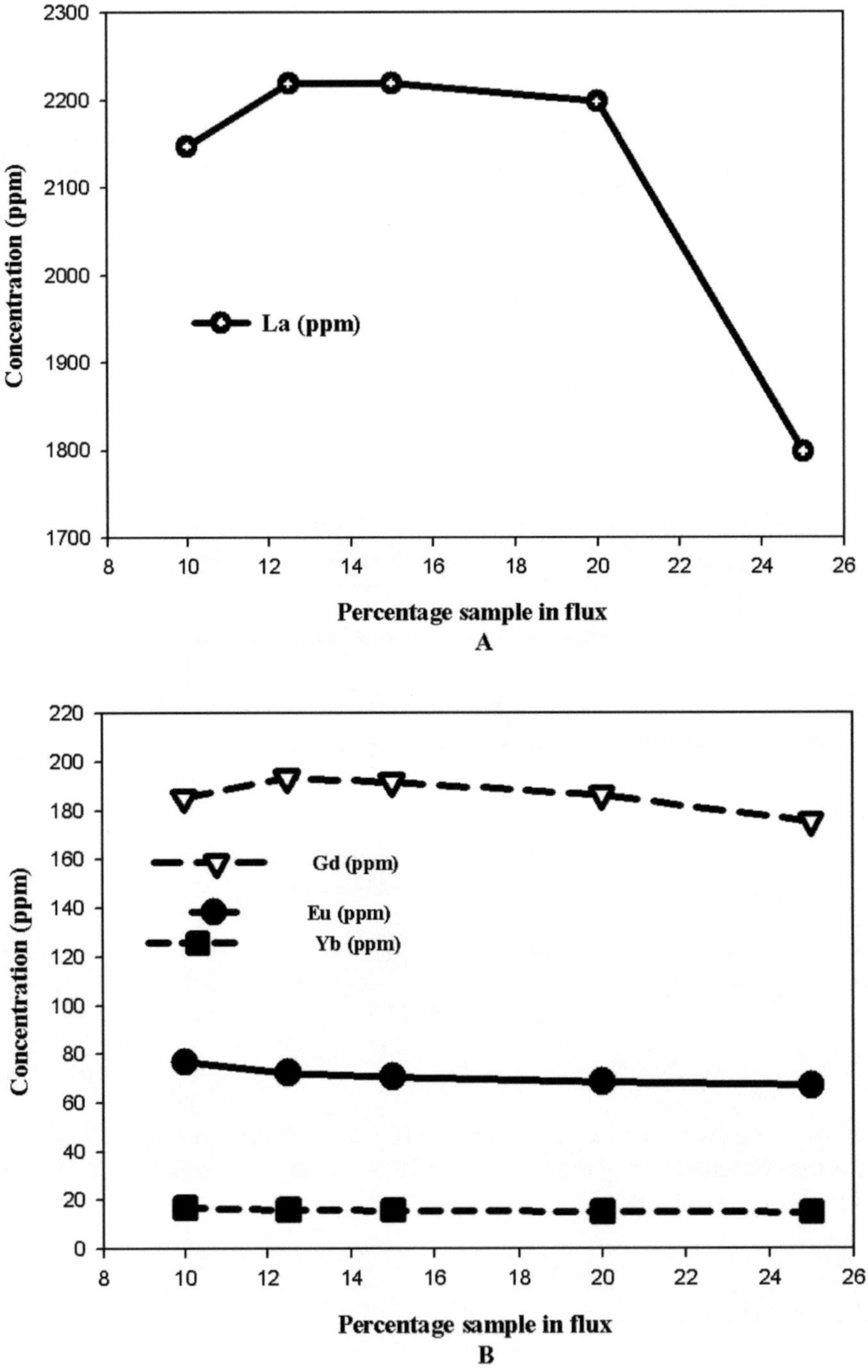

Figure 2. The effect of percentage of sample in flux (sample/flux ratio) on the concentration of La(II) (A) and Gd(II) and Eu(II) and Yb(II) (B) on GRE-03 reference material.

from 1.00 to 3.00 g. In most fusion techniques, a ratio of 1:10 sample/flux mixture is considered optimum for making fused beads. The best concentration of the reference material at different sample/flux ratio for lanthanum is 12% or 1:8 ratio. In **Figure 2(A)**, the results are showed for lanthanum and 2(B) for gadolinium, europium and ytterbium that were found to be best between 12 and 14%. High sample flux ratio may cause the sample not to be fully dissolved especially with the presence of refractive elements [16]. The flux content should be enough to fully dissolve the sample, but care is needed not to have too much of the flux as it will increase deposits on the ICP torch.

3.2. The effect of nebuliser gas flow rate

Figure 3 shows the results obtained for the behaviour of the four elements at selected wavelengths as a function of the nebuliser gas flow rate. The rare earth elements not only have extremely similar chemical properties but also have inner abundance differences among them that provide different responses to applied parameters [27].

The intensities of wavelength of the REEs, whether at first (5.3–6.2 eV) or second (10–12 eV) ionisation state, are all similar. The REEs possess soft ionic wavelengths allowing second ionisation potentials [28].

The classification of wavelengths is necessary as the effect of different operating parameters on analytical performance depends on them. Each nebuliser has got its own optimum gas flow that directly controls the sample uptake en route the plasma. The longer the sample interacts in the plasma, the more optical transitions of the elements are possible by it acquir-

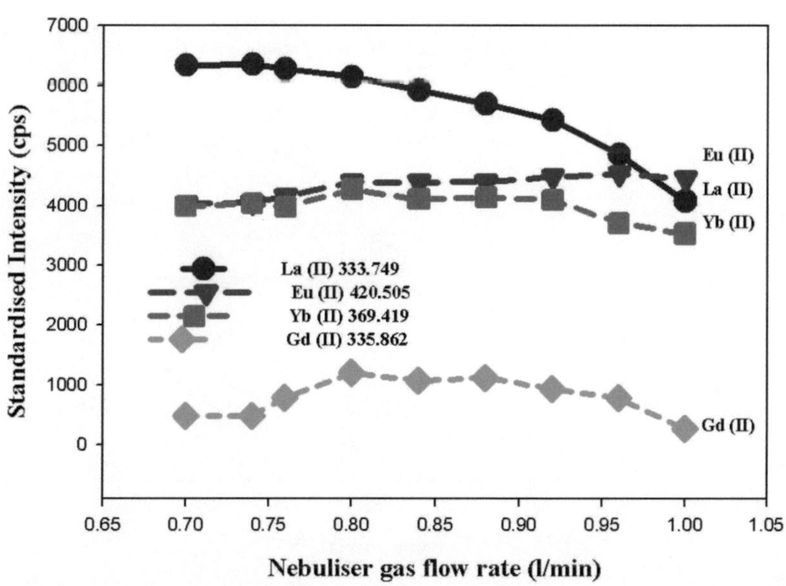

Figure 3. The effect of nebuliser gas flow on 0.1 mg L^{-1} of La(II), Yb(II), Eu(II) and Gd(II) standardised intensity counts, with the plasma power kept at 1300 W and pump speed at 35 rpm.

ing the energy for high energy interactions [29]. High nebuliser gas flow rate gives rise to reduced plasma temperature and lowers the atomisation. The low concentration of 0.1 mg L^{-1} of the REEs in $LiBO_3$–10% v/v HNO_3 was used for optimisation purposes.

The majority of the REEs studied to date gave best intensities at flow rates less than 0.82 L/min with La(II) [333.749] the lowest at 0.74 L/min [17]. An average flow rate at 0.80 L/min was chosen as the optimum nebuliser condition for all the rare earth elements studied.

3.3. The effect of plasma power

Power levels are critical when establishing optimum operating conditions. Higher power generates high temperature plasma conditions that lead to increases in intensity for atomisation and ionisation. The plasma power can have a major effect on the formation of oxides of lanthanides as well as in emission intensity and plasma robustness. Robust conditions of the plasma have been associated to high applied power by radiofrequency (RF) generator. **Figure 4** shows the results obtained for intensity measurements as a function of the RF power level for the four lines: La(II) [333.749], Eu(II) [420.505], Gd(II) [335.862] and Yb(II) [369.419] [17].

The standardised intensity rises with increased power levels due to the high energy of the plasma at higher power levels. The high plasma power also increases the background levels and sometimes at a rate higher than what the analyte increases. This rise indicates that the monitoring of background signal is important. On that instance, the best signal to background ratio needs to be established to get the optimum analyte signal and stable plasma. The standardised intensities rise with increase in RF power as shown in **Figure 4**. The plasma power

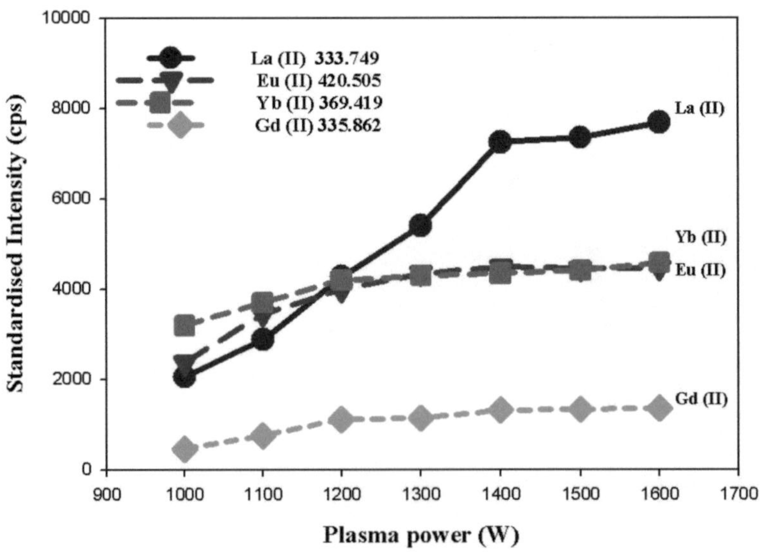

Figure 4. The effect of applied plasma power on 0.5 mg L^{-1} of La(II), Yb(II), Eu(II) and Gd(II) standardised intensity counts. The nebuliser gas flow rate was kept at 0.80 l/min and pump speed at 35 rpm.

of 1400 W was chosen as the optimum condition for the analysis of trace rare earth elements in a salt matrix of lithium metaborate [17].

3.4. The effect of pump speed

The pump speed affects the uptake of sample and the efficiency of nebulisation that is very critical to sensitivity. The nebuliser gas pressure and the speed of the peristaltic pump determine the volume uptake of sample and both influence the sample transit to the plasma. Large sample volumes increase the background level due to poor aerosol formation in the spray chamber.

The type and dimensions of pump tubing have effect on the pumping speed; hence, the ideal tubing must be sought [9]. Lower pump speed or using narrow bore pump tubing will reduce uptake rate that is better for high %TDS samples and suitable for fusion samples. The ideal tubing should be resistant to the solvent in use and to withstand low to high acid concentrations.

Figure 5 shows the behaviour of the standardised intensity for the four elements, as a function of pump speed. The experimental results shown in **Figure 5** exhibit the increase in optical transitions of La(II), Yb(II) and Eu(II) with increase in pump speed up to about 35 rpm, and then a decline was observed. Gd(II) did not show many variations due to the pump speed. Once the optimum pump speed has been determined for a specific nebuliser and sample matrix, it does not have to be changed on an element-by-element basis. The pump speed of 35 rpm was considered optimum in this study. However, upon changing the sample matrix and type of nebuliser, the pump speed must be verified.

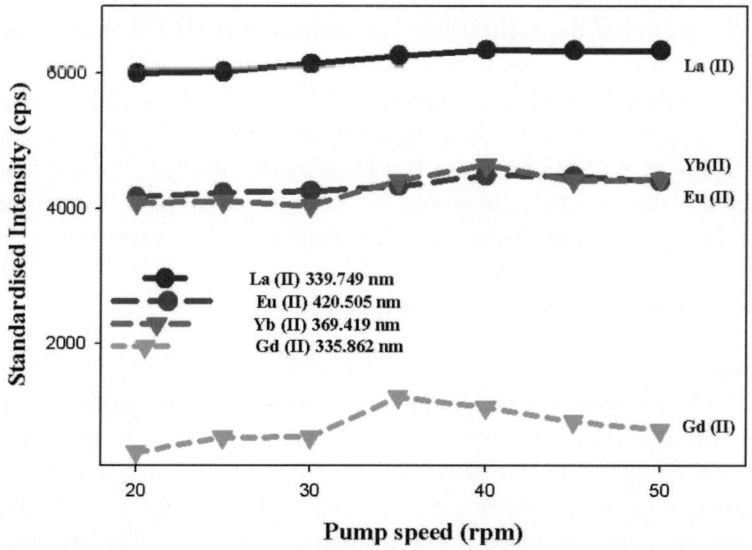

Figure 5. The effect of pump speed on 0.5 mg L^{-1} of La(II), Yb(II), Eu(II) and Gd(II) standardised intensity counts. The nebuliser gas flow rate was kept at 0.80 l/min and plasma power at 1300 W.

3.5. REEs optical transition

The REEs being a unique cluster of elements of atomic number from 57 to 71 increase in atomic number with a smooth decrease in ionic radius. This occurrence called lanthanide contraction applicable to all trivalent (triple positive charge) atoms makes the chemical properties of REEs very similar and hence they hold similar optical transitions [30]. The REEs have localised electrons that have interesting properties that originate from intraconfigurational transitions within the 4f level [31]. Due to high temperatures of the plasma (8000–10,000°C), there is a complete energy transfer in the form of emission and scattering of electromagnetic radiation. Er^{3+} and Yb^{3+} are examples of heavy rare earth elements with weak electron-phonon coupling, hence making emission simple because of the low energy involved and stable signals produced. The optimised conditions discussed in this chapter showed that different conditions give rise to variant optical transitions of the REEs as shown on plasma power of 1000–1600 W in **Table 3**. Detailed transitions of the REEs were outside the scope of this chapter.

3.6. Comparison of results in sediment samples

Two sediment samples, labelled VAD-01 and VAD-02 both collected from sampling sites near Durban harbour in South Africa, were taken for ICP spectroscopic analysis and the results are presented in **Table 3**.

3.6.1. Samples and reference material

The ICP results for two sediment samples, labelled VAD-01 and VAD-02, are presented in **Table 3**. The certified reference material GRE-03, a carbonatite from Tanzania rich in rare earth metals, was used to test the accuracy and recoveries. It should be noted that the matrix of this reference material does not fully match the sediments since we did not find a suitable sediment reference material with REEs. However, the reference material provided satisfactory results for the comparison of the three sample preparation techniques, fusion ICP-OES, acid ICP-OES and fusion ICP-MS analysis. The results obtained for the sample preparation technique provided further evidence that the different sample preparation techniques investigated were within the confidence limits of the results for reference material.

3.6.2. The comparison of results for two digestive methods investigated

The results obtained for the lithium metaborate fusion and four-acid digestion (HCl, HF, HNO_3 and $HClO_3$) procedures are expressed as the mean of three replicates as presented in **Table 3**. The solutions from both methods were analysed using the same conditions of the optimised ICP-OES procedure to determine the efficiency of the methods. The lithium metaborate digested samples were analysed on both ICP-OES and ICP-MS techniques to check the effectiveness of the optimised ICP-OES analysis conditions. The results of the heavy rare earth elements in **Table 3** from Sm to Lu clearly observed the Oddo-Harkins rule, which states that elements with even atomic number are more abundant than the elements with odd atomic number.

http://dx.doi.org/10.5772/intechopen.68280

Element (ppm)	Wavelength (nm)	ICP-OES (Fus)	ICP-OES (4-acids) [VAD-01]	ICP-MS	ICP-OES (Fus)	ICP-OES (4-acids) [VAD-02]	ICP-MS	ICP-OES (Fus)	ICP-OES (4-Acids)	ICP-MS	Ref value (95% confidence interval)
La	333.749	26.2	23.4	25.8	31.3	27.3	30.9	2216.7	2078.3	2218.6	2224 ±36
Ce	418.660	40.5	35.8	38.7	65.1	56.7	59.8	4352.9	4236.4	4361.8	4354 ±29
Pr	417.939	10.9	10.5	11.2	7.2	6.6	7.1	490.3	414.3	488.7	496.6 ±8.9
Nd	430.358	1.2	1.0	0.7	31.5	32.8	30.9	1855.9	1847.3	1856.1	1835.9 ±33
Sm	360.428	10.7	9.8	8.1	7.0	5.8	7.3	269.7	251.0	276.8	279.4 ±4.9
Eu	420.505	0.8	ND	0.3	1.6	1.7	1.6	72.1	68.3	71.9	75.24 ±2.0
Gd	335.862	8.3	5.6	9.4	7.9	7.4	7.7	196.6	161.2	193.4	191 ±3.6
Tb	321.890	1.1	0.6	1.5	2.2	2.1	1.8	20.4	13.2	20.2	21.65 ±0.42
Dy	338.502	5.5	3.8	5.4	8.2	8.2	7.6	92.9	69.7	89.9	92.33 ±2.2
Ho	345.600	0.8	ND	0.6	0.7	0.4	0.8	12.1	5.1	12.4	13.53 ±0.33
Er	349.910	2.3	1.6	2.4	1.9	1.5	0.7	11.3	10.9	16.8	28.84 ± 0.52
Tm	313.126	ND	ND	0.1	0.2	ND	0.1	2.7	2.6	2.9	3.08 ± 0.1
Yb	369.419	2.3	2.0	1.9	4.9	4.2	5.1	16.1	13.5	14.8	15.5 ±0.24
Lu	261.542	0.2	ND	0.4	ND	ND	0.2	0.9	ND	1.3	1.81 ±0.1
Y	371.030	16.3	16.6	18.0	24.1	22.8	25.1	319.5	311.9	317.9	320.6 ±5.7

Table 3. Comparison of analytical results obtained for spectroscopic evaluation by fusion ICP-OES, acid ICP-OES and fusion ICP-MS analysis of sediment samples.

The results obtained for the two samples and a reference material GRE-03 show enhanced results on most elements for the fusion method compared to the mineral acid digestion method. The acid digestion method does not completely dissolve some of the very resistant minerals and REEs, especially in hard rocks. Although the results are so close to that obtained in the fusion method, this can be attributed to the easy dissolution of the sediment material. The results suggest that the lithium metaborate fusion results in total dissolution of metals and is ideal for litho-geochemistry, including major oxides and trace rare earth metals. The detection limit of the acid digestion method was also compromised as we did not detect Eu, Ho and Lu in sample VAD-01, Tm in sample VAD-02 and Lu in sample GRE-03, but results were obtained for the fusion method. Generally, the acid digestion results for the GRE-03 sample was found to be lower than the certified values because of the difficulty in breaking the carbonatite in the sample matrix.

3.6.3. The comparison between ICP-OES and ICP-MS

The analytical techniques were performed using the same solution from the lithium metaborate fusion technique. The ICP-MS technique was found to be more efficient though costly in the trace elemental analysis, when compared to the ICP-OES technique. However, when comparing the results of ICP-MS with the optimised ICP-OES conditions, the former technique was very useful in validating the sediment results.

Most elements showed good agreement when the results are compared with less than 5% difference on all the samples analysed, including the reference material. For some elements with low levels present, such as Gd, Dy, Ho and Yb, the ICP-OES occasionally got slightly higher values than the ICP-MS technique. This observation in results was not significantly high, but it can be attributed to some spectral interferences.

The results obtained for the GRE-03 reference material compared well with the reported values listed in the table although the results for Ho, Tm and Lu were all moderately low on both instruments.

4. Conclusions

In this study, sample to flux ratio, optimum emission wavelengths, nebuliser flow rate, plasma power and pump speed were selected as major parameters to produce an analytical protocol for determining REEs in sediments. Of the two digestion procedures attempted, the more successful is obviously the flux-fusion method due to its ability of complete dissolution and simplicity due to automated fluxer fusion machine. The lower limit of detection was used as the optimisation criterion. The optimum values determined for sample to flux ratio, nebuliser gas flow rate, plasma power and pump speed (aspiration rate) were 1:8, 0.8 L min^{-1}, 1400 W and 35 rpm, respectively. The nebuliser gas flow was the most important parameter in the optimisation of the signal intensities. After optimisation of the parameters, the performance characteristics of the proposed method were established: linearity, detection and quantification limits and accuracy (recovery percentage), with and without addition of an internal standard.

http://dx.doi.org/10.5772/intechopen.68280

Acknowledgements

Special thanks for the support provided by the research laboratory at Scientific Services consulting analytical laboratory (Cape Town). The authors also like to thank the National Research Foundation (South Africa) and the CSIR (NRE, Stellenbosch) for supporting this research. We further acknowledge the support of the SensorLab, Chemistry Department, University of the Western Cape, Bellville, South Africa. A word of thanks also to the Chemistry Department, Faculty of Applied Sciences, Cape Peninsula University of Technology, Bellville, South Africa.

Author details

Martin Makombe[1], Charlton van der Horst[2], Bongiwe Silwana[2,3], Emmanuel Iwuoha[2] and Vernon Somerset[4,*]

*Address all correspondence to: somersetv@cput.ac.za

1 Scientific Services Consulting Analytical Laboratory, Cape Town, South Africa

2 SensorLab, Department of Chemistry, University of the Western Cape, Bellville, South Africa

3 Department of Chemistry, Durham University, Durham, United Kingdom

4 Department of Chemistry, Faculty of Applied Sciences, Cape Peninsula University of Technology, Bellville, South Africa

References

[1] Chaillou G, Anschutz P, Lavaux G, Blanc G. Rare earth elements in modern sediments of the Bay of Biscay (France). Marine Chemistry. 2006;**100**:39-52.

[2] Ali SH. Social and environmental impact of the rare-earth industries. Resources. 2014; **3**(1):123-134.

[3] Makombe M, Van der Horst C, Silwana B, Iwuoha E, Somerset V. Antimony film sensor for sensitive rare earth metal analysis in environmental samples. Journal of Environmental Science and Health A. 2016;**51**(8):597-606.

[4] Zawisza B, Pytlakowska K, Feist B, Polowniak M, Kita A, Sitko R. Determination of rare earth elements by spectroscopic techniques: A review. Journal of Analytical Atomic Spectrometry. 2011;**26**:2373-2390.

[5] Hall GM, Pelchat JC, Loop J. Determination of zirconium, niobium, hafnium and tantalum at low levels in geological materials by inductively coupled plasma mass spectrometry. Journal of Analytical Atomic Spectrometry. 1990;**5**:339-349.

[6] Rivoldini A, Fadda S. Inductively coupled plasma mass spectrometric determination of low-level rare earth elements in rocks using potassium-based fluxes for sample decomposition. Journal of Analytical Atomic Spectrometry. 1994;**9**:519-524.

[7] Lotter SJ, Purcelll W, Nel JT, Snyders E. Alternative dissolution of zircon samples and simultaneous analysis of major and trace components. The Journal of the Southern African Institute of Mining and Metallurgy. 2012;**112**:69-76.

[8] Amaral CDB, Machado RC, Barros JAVA, Virgilio A, Daniela Schiavo D, Nogueira ARA, Nóbrega JA. Determination of rare earth elements in geological samples using the Agilent SVDV ICP-OES [Internet]. 2016. Available from: www.agilent/cs/library/applications/5991-692EN.pdf [Accessed: 21 October 2016].

[9] Li JX, Zhu ZW, Yin XF, Han B, Zheng L, Wang JT, Wang XR. Analysis of contents and distribution patterns of rare earth elements in the surface sediments of the south mid-Atlantic ridge. Chinese Journal of Analytical Chemistry. 2015;**43**(1):21-26.

[10] Bayon G, Barrat JA, Etoubleau J, Benoit M, Bollinger C, Revillon S. Determination of rare earth elements, Sc, Y, Zr, Ba, Hf and Th in geological samples by ICP-MS after Tm addition and alkaline fusion. Geostandards and Geoanalytical Research. 2009;**33**(1): 51-62.

[11] Claisse. Claisse M4 gas fusion user manual. Corporation Scientific Claisse Inc., Zurich, Switzerland 2010; 27-38

[12] Jin X, Zhu H. Determination of platinum group elements and gold in geological samples with ICP-MS using a sodium peroxide fusion and tellurium co-precipitation. Journal of Analytical Atomic Spectrometry. 2000;**15**:747-751.

[13] Obaje SO, Akpoborie IA, Ugbe FC, Onugba A. Rare earth and trace elements distribution in sediments of river Gora, Minna area, North-Central Nigeria: Implication of provenance. Earth Science Research. 2015;**4**(1):103-112.

[14] Zhang Y, Gao X, Chen CTA. Rare earth elements in intertidal sediments of Bohai Bay, China: Concentration, fractionation and the influence of sediment texture. Ecotoxicology and Environmental Safety. 2014;**105**:72-79.

[15] Panteeva SV, Gladkochoub DP, Donskaya TV, Markova VV, Sandimirova GP. Determination of 24 trace elements in felsic rocks by inductively coupled plasma mass spectrometry after lithium metaborate fusion. Spectrochimica Acta B. 2003;**58**:341-350.

[16] Ramteke DD, Annapurna K, Deshpande VK, Gedam RS. Effect of Nd^{3+} on spectroscopic properties of lithium borate glasses. Journal of Rare Earths. 2014;**32**(12):1148-1153.

[17] Bentlin FRS, Pozebon D. Direct determination of lanthanides in environmental samples using ultrasonic nebulization and ICP OES. Journal of Brazilian Chemical Society. 2010;**21**(4):627-634.

[18] Spectro analytical instruments [Internet]. 2016. Available from: www.spectro.com [Accessed: 11 November 2016].

[19] Agilent Technologies. AutoMax Fast automated method optimization, Technical Overview 700 Series ICP OES [Internet]. 2010. Available from: www.agilent.com/cs/library/technical overviews/public/si-0286.pdf [Accessed: 21 October 2016].

[20] Silva FV, Trevizana LC, Cintia Silva S, Nogueira Ana Rita A, Nobrega JA. Valuation of inductively coupled plasma optical emission spectrometers with axially and radially viewed configurations. Spectrochimica Acta B. 2002;**57**:1905-1913.

[21] Guimarães-Silva AK, de Lena JC, Froes RES, Costa LM, Nascentes CC. Evaluation of signal-to-background and Mg II/Mg I ratios as response for the optimization of rare earth elements determination by inductively coupled plasma optical emission spectrometry. Journal of Brazilian Chemical Society. 2012;23(4):753-762.

[22] Hou X, Jones BT. Inductive coupled plasma/optical emission spectrometry. Encyclopedia of Analytical chemistry, John Wiley and Sons Ltd, Chichester. 2000;9468-9485.

[23] Santos FN, Silva IS, Gomide RG, Oliveira LC. Evaluation of the rare earth impurities determination in Gd_2O_3 matrix by ICP-OES. In: International Nuclear Atlantic Conference – INAC; 24-29 November 2013; Recife, PE, Brazil. Associação Brasileira De Energia Nuclear – ABEN. ISBN: 978-85-99141-05-2.

[24] Bourmans PWJM. Measuring detection limits in inductively coupled plasma emission spectrometry using SBR-RSDB approach: A tutorial discussion of theory. Spectrochimica Acta. 1991;**46B**:431-445.

[25] Bourmans PWJM, Tielrooy JA, Muessen F. Mutual spectral interferences of rare earth elements in inductively coupled plasma atomic emission spectrometry: Rational line selection and correction procedure. Spectrochimica Acta. 1988;**43B**:173-199.

[26] Sturman BT, Doidge PS. Semi empirical approach to the calculation of instrumental detection limits in inductively coupled plasma atomic emission spectrometry. Journal of Analytical Atomic Spectrometry. 1998;**13**:69-74.

[27] Wu S, He M, Hu B, Jiang Z. Determination of trace rare earth elements in natural water by electrothermal vaporisation ICP-MS with pivaloyltrifluoroacetone as chemical modifier. Microchimica Acta. 2007;**159**:269-275.

[28] Barbalace K. Periodic table of elements sorted by 1st ionization potential(eV) [Internet]. 1995. Available from: www.environmentalchemistry.com/yogi/periodic/1stionization.html [Accessed: 28 October 2016].

[29] Boumans PWJM. Inductive coupled plasma emission spectroscopy, part1. Methodology, instrumentation and performance, John Wiley and Sons. New York, USA. 1987

[30] Castor SB, Hedrick JB. Rare earth elements. In: Kogel JE, Trivedi NC, Barker JM, Krukowski ST, editors. Industrial minerals and rocks – commodities, markets and uses. 7th ed. Society for Mining, Metallurgy and Exploration, Inc. (SME). Littleton, Colo. 2006. pp. 769-792.

[31] Knyazev Yu V, Noskov MM. The optical properties of rare earth metals. Physica Status Solidi (B). 1977;**80**:11.

Chapter 4

Theoretical Spectroscopy of Rare-Earth Elements: Spectra and Autoionization Resonances

Andrey A. Svinarenko, Alexander V. Glushkov,
Olga Yu. Khetselius, Valentin B. Ternovsky,
Yuliya V. Dubrovskaya, Anna A. Kuznetsova and
Vasily V. Buyadzhi

Additional information is available at the end of the chapter

http://dx.doi.org/10.5772/intechopen.69314

Abstract

An investigation of spectra, radiative and autoionization characteristics for the rare-earth elements is of a great interest as for development atomic spectroscopy as different applications in plasma chemistry, astrophysics, laser physics, quantum electronics etc. We present and review the results of studying spectra and autoionization resonance characteristics of a few lanthanide elements (ytterbium and thulium). Computing the spectra and autoionization resonance parameters is carried out within the relativistic many-body perturbation theory (RMBPT) and generalized relativistic energy approach. The accurate results on the autoionization resonance energies and widths are presented with correct accounting for the exchange-correlation and relativistic corrections and compared with other available theoretical and experimental data. In this chapter, we present a brief review of the theoretical and experimental works on spectroscopy of some lanthanide atoms. Spectroscopy of the Rydberg autoionization resonances in rare-earth atoms in an external electromagnetic field is expected to be very complex and unusual.

Keywords: rare-earth elements, spectroscopy, relativistic many-body perturbation theory, spectra, autoionization resonances

1. Introduction

An investigation of spectra, optical and spectral, radiative and autoionization characteristics for the rare-earth elements (REEs) (isotopes) and corresponding ions is traditionally of a great interest due to the further development of quantum optics and atomic spectroscopy and their different applications in the plasma chemistry, astrophysics, laser physics, quantum and nanoelectronics, etc. (see Refs. [1–8]). As it was indicated in many information sources on the rare-earth elements, they make possible the high-tech world—everything from the miniaturization of electronics, to the enabling of green energy and medical technologies, to supporting a myriad of essential telecommunications and defence systems [1, 2]. Besides, it is worth to mention that such industrial operations such as mining, refining and recycling of rare earths can have serious environmental consequences in a case of the improper operations and management. As it was indicated in Ref. [1, 2] that the rare-earth elements all tend to occur together in the same mineral deposits with radioactive elements, particularly thorium and uranium, and because they have similar properties, it is difficult to separate them from one another. Let us remember that according to the modern IUPAC definition, the rare-earth element (REE) is one of a set of 17 chemical elements in the Mendeleev's periodic table, specifically the 15 lanthanides, as well as scandium and yttrium.

Below we will present the results of theoretical investigation of a few lanthanide atoms, namely, ytterbium and thulium. The detailed review of studying energetic, spectral and radiative characteristics of these atoms is presented in a number of monographies and articles [11, 14–86]. One should mention the detailed theoretical studying of the lanthanide atoms by Cowan and pioneer studying by King and Meggers et al., who studied atomic spectra of rare-earth elements and reported wavelengths and relative intensities of more than 1600 lines and estimated intensities of more than 1700 lines of Yb I (see [6–9] and references therein). Camus [14] and Wyart and Camus [15] measured more than 70 absorption lines in the ultraviolet region and obtained the even-parity states with $J = 0$ and $J = 2$ belonging to the series $4f^{14}6sns$ (n = 13–62), $4f^{14}6snd$ (n = 11–64) and $4f^{14}6snd$ (n = 11–21). Moreover, these authors fulfilled new additional observations of the neutral ytterbium spectrum in the infrared region and measured the Zeeman resonance parameters in the visible and near ultraviolet regions. In fact, these measurements allow to propose a new classification of the $4f^{14}$ $6snl$ Rydberg series in the Yb spectrum. The configurations $4f^{13}6s^26p$ and $4f^{14}6p^2$ have been complete and the lowest levels of $4f^{13}5d^26s$ and $4f^{14}5d6p$ have been recognized, as well as new low members of the Rydberg series. Spector [16] identified the energy levels of $4f^{13}5d6s^2$ and $4f^{13}6s^26p$ configurations. Highly excited levels of the neutral ytterbium have been studied by means of the two-photon and two-step spectroscopy of even spectra [17] and theoretically by means of the multi-channel quantum defect analysis of the f odd- and even-parity spectra [18].

In Ref. [19], it has been used the method of photoionization laser spectroscopy of the ytterbium atoms in the presence of a direct current (DC) electric field on order to study the Yb photoionization near the first ionization threshold. Blondel et al. [19] had recorded a few spectra corresponding to ionization from the levels $4f^{14}6s7s\ {}^1S_0$ and ${}^3S_1$, $4f^{14}6s6p\ {}^1P_1{}^0$ and $4f^{14}6s5d\ {}^1D_2$. The first observation of a long Rydberg series due to single excitation from the $4f^{14}$ subshell in Yb I was reported in Ref. [20] and additionally a two-channel multi-channel quantum defect analysis reveals a prominent series perturbation due to coupling with one particular $4f^{14}5dnf$ doubly excited series. Baig et al. [21] have studied the inner shell and double excitation spectrum of ytterbium involving the $4f$ and $6s$ subshells and reported more than 200 new levels. Theoretical analysis allowed to interpreter these lines as the $4f^{14}6s^2$ to $4f^{13}({}^2F_{712,512})nd,ng$ inner shell transitions and the $4f^{14}6s^2$ to $4f^{14}5d(^2D_{312,512})np,nf$ and $4f^{14}6s^2$ to $4f^{14}6p(^2P_{112,312})ns,nd$ doubly excited ones. All the observed levels lie above the first ionization threshold and can be ordered into Rydberg series converging onto six limits. The interchannel interactions between the overlapped series have been parametrized using multi-channel quantum defect theory.

In Ref. [22], Karaçoban and Özdem carried out the total relativistic computing the energies, Landé factors, lifetimes for a group of the excited levels outside the neutral ytterbium core [Xe]. These authors have used the relativistic Hartree-Fock (RHF) method (the Cowan's atomic code) and the multi-configuration Hartree-Fock (MCHF) method with accounting of the relativistic effects within the Breit-Pauli (BP) scheme (the Fischer's atomic code).

Maeda et al. [23] have used a method of optical-microwave double-resonance spectroscopy to study the highly excited Rydberg states of ytterbium. In Refs. [24, 25], the methods of two-photon ionization and three-photon polarization spectroscopy have been applied to studying the autoionization states (ASs) of the ytterbium atom. The ytterbium atom in the $6s^2\ ^1S_0$ ground state was excited to the 6s6p $^3P_1$ excited state by a photon with a visible wavelength (555.648 nm), and it was ionized by absorbing an ultraviolet photon with a wavelength of 260–285 nm. The Rydberg and autoionization states of neutral ytterbium are also considered in Ref. [27]. In Ref. [27], the authors reported the results of a reinvestigation of the 6snf $^{1,3}F_3$ ($9 < n < 30$), $6snp\ ^1P_1$ ($12 < n < 56$), $6snp\ ^3P_1$ ($12 < n < 21$) and $6snp\ ^3P_2$ ($12 < n < 19$) Yb Rydberg states using a two-colour three-photon resonant excitation technique through the $4f^{14}6s5d\ ^1D_2$ intermediate level. The novelty of this studying is in the identification of the new levels in the vicinity of the $6snf$ ($9 < n < 16$) Rydberg levels.

The laser spectroscopy method has been effectively applied by Letokhov [28, 29] and Letokhov and co-workers [30, 31] to studying narrow doubly excited autoionization states of ytterbium and other lanthanide atoms. An analysis is made of doubly excited autoionization states of an atom with two valence electrons. The results of experimental and theoretical investigations of narrow autoionization states are reported for ytterbium atoms near the ionization threshold. The method of multi-stage photoionization of atoms by tunable laser radiation was used to detect experimentally and identify narrow autoionization states $^1P_1^0$ and $^3P_{0,1,2}^0$ of the $7s6p$ configuration. In Refs. [32–48], it has been performed the detailed computing energies and widths of the autoionization resonances, Rydberg levels for ytterbium and thallium. The positions and widths of the autoionization states belonging to the $7s6p$, $6p5d$, $6p^2$ and $5d^2$ configurations were calculated using the method of relativistic perturbation theory (PT) with the model potential (MP) zeroth approximation [49–79]. In Refs. [63–68, 85, 86], it has been discovered a principally new spectroscopic effect of a giant broadening autoionization resonances of the lanthanide atoms (thulium and gadolinium) in a sufficiently weak external electric (laser) field. This new effect is of a great importance for problem of laser separation of heavy isotopes and nuclear isomers, spectroscopy and photochemistry [65, 66, 93, 94]. It is worth to note from theoretical viewpoint that computing spectra, radiative transition and autoionization parameters of the rare-earth elements' energy is very complicated task because of the necessity of the correct accounting for the exchange-correlation (including polarization and screening effects, a continuum pressure, etc.) and relativistic corrections (and also radiative and nuclear effects in a case of the super heavy atomic systems). Moreover, theoretical study of spectral properties of the lanthanide atoms is of a great importance for the development of new methods of atomic spectroscopy.

Let us remind that many different quantum-mechanical approaches in the on-relativistic and relativistic versions have been used in order to study energetic and spectroscopic parameters of the lanthanide elements. One should mention such known methods as a standard multi-configuration Hartree-Fock (HP) method with accounting of the relativistic effects within the Breit-Pauli scheme or the classical relativistic Hartree-Fock method or, at last, a multi-configuration Dirac-Fock method. These methods allow to obtain very useful spectral data about many light and heavy atoms. Nevertheless, studying the autoionization resonance characteristics for heavy elements with using the mentioned methods not always provides a precise description of the corresponding spectra. From the other side, at present time, the advanced versions of these methods have been developed, where the one- and two-particle relativistic and exchange-correlation effects are taken into account very precisely. It is worth to remind about computer codes for relativistic many-body calculations of atomic (molecular) properties developed in the Oxford group, Russian-German one, etc. ('GRASP', 'Dirac'; 'BERTHA', 'QED'; see Refs. [1–13] and references therein).

In Refs. [71–88], it has been developed a new formalism of the relativistic many-body PT with using the optimized one-quasiparticle (QP) representation and effective account of the exchange-correlation corrections of the PT second order and higher orders (polarization interaction, quasiparticles screening, etc.). The method of the relativistic many-body PT is constructed on the basis of the same ideas as the well-known PT with the model potential zeroth approximation by Ivanov et al. [49–64]. However, there are two key differences: namely, the PT zeroth approximation [81, 82, 85, 86] is in fact the Dirac-Kohn-Sham one. In order to calculate the radiative transition and autoionization resonance parameters, a new version of relativistic energy approach is used. It is important to remind that a model relativistic energy approach in a case of the multi-electron atom has been proposed by Ivanov et al. [49–56]. A generalized gauge-invariant relativistic energy approach in a case of the multi-electron atomic systems has been developed by Glushkov-Ivanov-Ivanova (see Refs. [71–76]). It should be noted that an energy approach uses the known Gell-Mann and Low S-matrix scheme. More exactly, a probability of any atomic state radiative decay is connected with an imaginary part of an electron energy shift ΔE, which can be expressed through the Quantum electrodynamics (QED) scattering matrix (by means of the Gell-Mann and Low formula), including the interaction with the photon vacuum field (the spontaneous radiative decays) as an external electromagnetic field (the induced decays) [67–75, 85, 86, 91 92]. In Refs. [71–76], the new energy approach has been combined with a formalism of the relativistic many-body perturbation theory (RMBPT) with an optimized one-quasiparticle representation and an accurate accounting of the multi-electron exchange-correlation effects.

In this chapter, we present the results of application the optimized relativistic many-body PT with the Dirac-Kohn-Sham zeroth approximation and a generalized relativistic energy approach to studying spectra, radiative and autoionization resonance characteristics for lanthanide atoms, in particular, ytterbium and thulium. Data on the autoionization resonance energies and widths in the ytterbium and thulium are obtained with correct accounting for the exchange-correlation and relativistic corrections and compared with other available theoretical and experimental data.

2. Method of relativistic many-body perturbation theory and relativistic energy approach

As our method of computing has been in detail presented previously, here we summarize only the key points. Generally speaking, the energy spectra for the majority of complex atomic systems (naturally including the rare-earth elements) are characterized by a great density. Moreover, these spectra have essentially relativistic properties. So, a correct theoretical method of their studying can be based on the convenient field procedure, which includes computing the energy shifts ΔE of the degenerate electron states. More exactly, speech is about constructing secular matrix M (with using the Gell-Mann and Low adiabatic formula for ΔE), which is already complex in the relativistic theory, and its further diagonalization [49, 50, 55, 56]. In result, one could compute the energies and decay probabilities of a non-degenerate excited state for a complex atomic system [49–64]. The secular matrix elements can be further expanded into a PT series on the interelectron interaction. Here, the standard Feynman diagrammatic technique is usually used.

Generally speaking, the secular matrix M can be represented as follows:

$$M = M^{(0)} + M^{(1)} + M^{(2)} + M^{(3)}. \quad (1)$$

where $M^{(0)}$ is the contribution of the vacuum diagrams of all PT orders (this contribution determines only the general levels spectrum shift); $M^{(1)}$, $M^{(2)}$ and $M^{(3)}$ are contributions of the 1-, 2- and 3-quasiparticle (QP) diagrams, respectively. The matrix $M^{(1)}$ can be presented as a sum of the independent one-QP contributions. Substituting these quantities into Eq. (1) one could have summarized all the one-QP diagram contributions. In the empirical methods here, one could use the experimental values of one-electron energies; however, the necessary experimental quantities (especially for the rare-earth and other elements) are not often available. The detailed procedure for computing Re $M^{(2)}$ is presented, for example, in Refs. [51–56].

We will describe an atomic multi-electron system by the relativistic Dirac Hamiltonian (the atomic units are used) as follows [17, 18]:

$$H = \sum_i \{\alpha c p_i - \beta c^2 - Z/r_i\} + \sum_{i>j} \exp(i|\omega|r_{ij})(1 - \alpha_i\alpha_j)/r_{ij} \quad (2)$$

where Z is a charge of nucleus, α_i and α_j are the Dirac matrices, ω_{ij} is the transition frequency, c is the velocity of light. The interelectron interaction potential (the second term in Eq. (3)) takes into account the retarding effect and magnetic interaction in the lowest order on the parameter of the fine structure constant. In the PT zeroth approximation, it is used ab initio mean-field potential:

$$V^{DKS}(r) = [V_{Coul}^{D}(r) + V_X(r) + V_C(r|b)] \tag{3}$$

with the standard Coulomb, exchange Kohn-Sham V_X and correlation Lundqvist-Gunnarsson Vc potentials (see details in Refs. [75–79]). An effective approach to accounting the multi-electron polarization contributions is described earlier and based on using the effective two-QP polarizable operator, which is included into the PT first-order matrix elements.

In order to calculate the radiation decay probabilities and autoionization energies and widths, a gauge invariant relativistic energy approach (version [71–76]) is used. In particular, a width of the state, connected with autoionization decay, is determined by coupling with the continuum states and calculated as square of the matrix element [73–76]:

$$V_{\beta_1\beta_2;\beta_4\beta_3} = \sqrt{(2j_1+1)(2j_2+1)(2j_3+1)(2j_4+1)}(-1)^{j_1+j_2+j_3+j_4+m_1+m_2} \times \sum_{a\mu}(-1)^{\mu}\begin{pmatrix} j_1 & j_3 & a \\ m_1 & -m_3 & \mu \end{pmatrix}\begin{pmatrix} j_2 & j_4 & a \\ m_2 & -m_4 & \mu \end{pmatrix} \times Q_a(n_1l_1j_1n_2l_2j_2; n_4l_4j_4n_3l_3j_3) \tag{4}$$

Here $Q_a = Q_a^{Qul} + Q_a^{Br}$, where Q_a^{Qul} and Q_a^{Br} correspond to the Coulomb and Breit parts of the relativistic interelectron potential in Eq. (3) and express through Slater-like radial integrals and standard angle coefficients. Other details can be found in Refs. [71–92].

The most complicated problem of the relativistic PT computing the rare-earth element spectra is in an accurate, precise accounting for the multi-electron exchange-correlation effects (including polarization and screening effects, a continuum pressure, etc.), which can be treated as the effects of the PT second and higher orders. Using the standard Feynman diagrammatic technique, one should consider two kinds of diagrams (the polarization and ladder ones), which describe the polarization and screening exchange-correlation effects. The detailed description of the polarization diagrams and the corresponding analytical expressions for matrix elements of the polarization QPs interaction (through the polarizable core) potential is presented in Refs. [51–56]. An effective approach to accounting of the polarization diagram contributions is in adding the effective two-QP polarizable operator into the PT first-order matrix elements. In Refs. [55, 56], the corresponding non-relativistic polarization functional has been derived. More correct relativistic expression has been presented in Refs. [11, 81, 82] and used in our computing. The corresponding polarization potential looks as follows [81, 82]:

$$V_{pol}^{d}(r_1r_2) = X\left\{\int \frac{dr'\left(\rho_c^{(0)}(r')\right)^{1/3}\theta(r')}{|r_1-r'|\cdot|r'-r_2|} - \int \frac{dr'\left(\rho_c^{(0)}(r')\right)^{1/3}\theta(r')}{|r_1-r'|}\int \frac{dr''\left(\rho_c^{(0)}(r'')\right)^{1/3}\theta(r'')}{|r''-r_2|} \Big/ \left\langle\left(\rho_c^{(0)}\right)^{1/3}\right\rangle\right\} \tag{5a}$$

$$\left\langle\left(\rho_c^{(0)}\right)^{1/3}\right\rangle = \int dr\left(\rho_c^{(0)}(r)\right)^{1/3}\theta(r), \tag{5b}$$

$$\theta(r) = \left\{1 + [3\pi^2 \cdot \rho_c^{(0)}(r)]^{2/3}/c^2\right\}^{1/2} \tag{5c}$$

where ρ_c^0 is the core electron density (without account for the quasiparticle), X is numerical coefficient and c is the light velocity. The contribution of the ladder diagrams (these diagrams describe the immediate QPs interaction) is summarized by a modification of the PT zeroth approximation mean-field central potential (see below), which include the screening (anti-screening) of the core potential of each particle by the two others. The details of this contribution can be found in Refs. [71–92]. All computing were performed with using the modified PC code 'Superatom-ISAN'.

3. Spectra and autoionization state characteristics for the ytterbium and thulium atoms

3.1. Spectroscopy of excited and autoionization states in ytterbium

The Yb ($Z = 70$) atom has the external electron configuration $4f^{14}6s^2$ in the ground state and any excited states of the ytterbium atom can be treated as the states with two- or three quasiparticles above the electron core [Xe]$4f^{14}$. **Figure 1** shows a qualitative spectrum of the ytterbium, according to Ref. [30, 31].

In the Yb spectrum, it is possible to define two main types of the autoionization states (ASs) [28, 29]. The state of the first type arises with the excitation of inner shells, the condition of the second type with the double excitation of the valence shell. Those arising from the excitation of the 4*f*-shell belong to the intermediate type. Autoionization decay in single-particle approximation can be represented as follows: $\alpha_1\alpha_2 \rightarrow \alpha_3 k$, where α_i (i = 1, 2, 3) describes the set of quantum numbers of bound states, k is the state of free electron. The decay is possible only in a state of continuum, which matches the original AS by parity and value of a total momentum J.

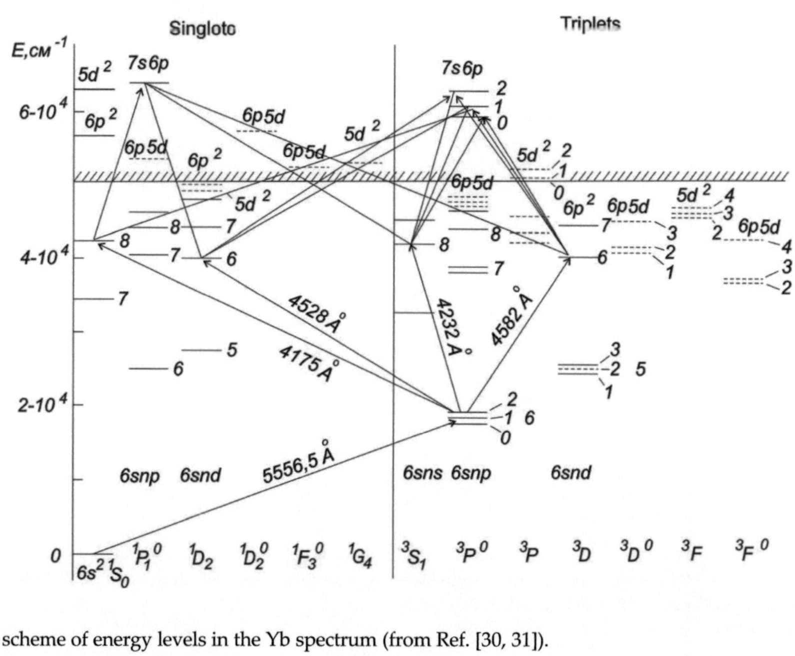

Figure 1. The scheme of energy levels in the Yb spectrum (from Ref. [30, 31]).

Low-lying AS of the Yb atom, arising from the double excitation of $6s^2$ shell, decays under the scheme $n_1l_1n_2l_2 \rightarrow 6skl$. State with electronic configuration $4f^{13}6s^27nl$ is a low-lying autoionization state AS with excitation of electron-hole type. The energy areas of these states and states with excited $6s^2$-shell match. The decay of such AS takes place under the scheme: $4f^{13}6s^27nl \rightarrow 4f^{14}6skl'$. Wide low-lying AS ($\Gamma \approx 100$ cm^{-1}) observed in the experiment by Letokhov and co-workers [30, 31] are attributed to such states. The estimation of the AS energy with configurations *6p5d*, $6p^2$, $5d^2$, *7s6p* in one-particle approximation indicates that these states should be located in the area of 0–15,000 cm^{-1} above Yb ionization threshold. As one could see from **Figure 1**, it is clear that in this case, even configurations $6p^2$, $5d^2$ can be excited in two stages, and the odd *6p5d*, *7s6p* can be excited in three stages. This fact has been used in the experiments by Letokhov and co-workers [30, 31]. In this experiment, the Yb atoms were excited into the AS by radiation of three dye lasers. The laser of the first stage (λ_1 = 5556.5 Å) excited transition $6s^{21}S_0 \rightarrow 6s6p\ P_1^0$ (see **Figure 1**). The second laser provided the further excitation to the one of selected intermediate states. The wavelength of the third laser was rebuilt in a such range it provided AS excitation in the studied area 0–15,000 cm^{-1} above the ionization threshold. In particular, upon AS excitation through $6s6d^1D_2$ state the wavelength of the third stage varied in the range 7100–4100 Å, which allowed to ionize Yb atoms, excited in $^1D_2$ state and carry out search of AS in the area of 3700–14,000 cm^{-1}. **Figure 2** shows the dependence of ion current on the third stage laser wavelength, under excitation of the AS *7s6p* $^3P0^0$ from the $6s6d^3D_1$ state by laser radiation with the line width of ~1 cm^{-1}. **Figure 2** also shows the AS $^3P0^0$ with the width of the laser line of the three stage ~0.1 cm^{-1}.

In **Table 1**, we present the experimental [51, 54] and theoretical data for the energy (energy count from ground state $4f^{14}6s^2\ ^1S_0$) of some YbI singly excited states: MCHF-BP, data obtained on the basis of multi-configuration Hartree-Fock (MCHF) method with Breit-Pauli (BP) adjustments (A, B + D, D different sets of configurations included in the calculation by the method MCHF-BP [22]); RHF—the relativistic Hartree-Fock data [1, 2]; DF—analysis of Wyart-Camus

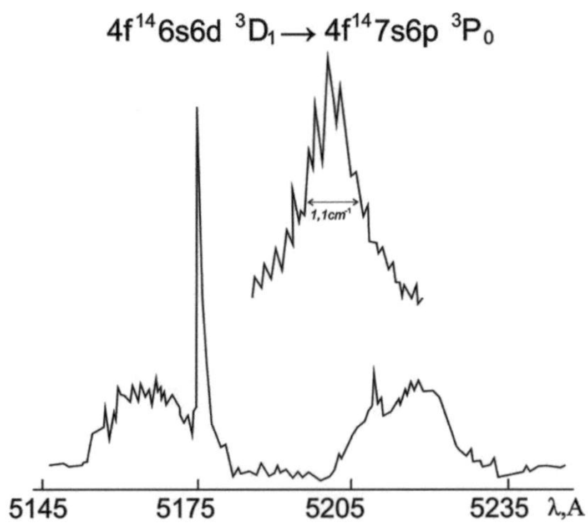

Figure 2. Dependence of ion current on the third stage laser wavelength, under excitation of the AS *7s6p* $^3P0^0$ from the $6s6d^3D_1$ state by laser radiation with the line width of ~1 cm^{-1}.

Config.	J	MCHF + BP(A)	MCHF + BP(C)	MCHF + BP(BD)	RHF	RPTMP	Our theory	DF	Exp
$6s_{1/2}{}^{2*}$	0	0	0	0	0	0	0	0	0
$6s_{1/2}6p_{1/2}$	0	18,087	17,262	18,730	17,320	17,400	17,310	17,312	17,288
$6s_{1/2}6p_{1/2}$	1	18,174	17,568	18,813	17,954	18,100	18,008	17,962	17,992
$6s_{1/2}6p_{3/2}$	1	24,614	26,667	25,257	25,069	25,500	25,094	25,075	25,068
$6s_{1/2}6p_{3/2}$	2	18,357	18,249	18,999	19,710	19,800	19,715	19,716	19,710
$6s_{1/2}5d_{3/2}$	1	24,094	28,871	23,740	24,489	23,900	24,410	24,489	24,489
$6s_{1/2}5d_{3/2}$	2	24,505	28,973	24,172	24,484	24,600	24,824	24,751	24,752
$6s_{1/2}5d_{5/2}$	2	26,984	29,633	26,841	27,677	26,100	26,970	27,654	27,678
$6s_{1/2}5d_{5/2}$	3	25,860	29,374	25,500	25,271	24,900	25,098	25,270	25,271

* E = −14,8710 cm^{-1}; E1 = −148,700 cm^{-1}; E2 = −148,695 cm^{-1} [45–48].

Table 1. Energies (cm^{-1}) of YbI singly excited states: MCHF-BP, with Breit-Pauli (BP) adjustments (A, B + D, D, different sets of configurations); RHF, relativistic Hartree-Fock method; RPTMP (E1), relativistic perturbation theory with model potential (MP) zeroth approximation [30, 31]; DF, analysis of Wyart-Camus [15]; RPT-DKS (E2), our theory and experiment data (Exp).

[15]; RPTMP (E1)—relativistic perturbation theory with model potential (MP) zeroth approximation by Ivanov et al. [30, 31]; RPT-DKS (E2)—our theory and experiment data (Exp).

The analysis of the calculated data (**Table 1**) shows that the role of exchange-correlation effects for the studied atom is extremely significant. The HF method with a small number of considered configurations gives an error of more than 100 cm^{-1}. In **Table 2**, we list the experimental and theoretical data of Letokhov and co-workers [30, 31] for the energy and width of the excited (AS) states of the $7s6p$ configuration in the YbI spectrum (counted from the ground state $4f^{14}6s^{2}\,{}^{1}S_{0}$Yb): E1, Γ1—RMBPT data by Ivanov et al. [30, 31]; E2, Γ2—our theory (RPT-DKS); E3—MCHF-BP data by Karaçoban-Özdemir [22] (classification in [19] differs from our classification).

The analysis of data in **Table 2** shows that the values E1, E2 and E_{exp} are in good agreement with each other, however, values Γ1 and Γ_{exp} differ significantly. In our opinion, this fact is due to the lack of accurate estimates of the radial integrals, use of non-optimized bases and some other approximations of the calculation. This applies to data obtained on the basis of RHF and MCHF methods. In our calculation, we used optimized one-QP representation and optimal relativistic bases and accounted more accurately for the important multi-particle exchange-correlation effects, including polarization and screened interaction of quasiparticles, continuum

Term	Theory E3	Theory E1	Theory Γ1	Theory E2	Theory Γ2	Exp. E_{exp}	Exp. Γ_{exp}
${}^{3}P_{0}{}^{0}$	61,233	59,800	0.7	59,450	1.15	59130.5	1.1
${}^{3}P_{1}{}^{0}$	62,085	60,000	3.0	60,315	1.10	60428.7	0.95
${}^{3}P_{2}{}^{0}$	62,423	62,600	0.7	62,587	1.51	62529.1	1.6
${}^{1}P_{1}{}^{0}$	64,216	63,600	1.8	63,613	2.48	63655.8	2.6

Table 2. Energies E (cm^{-1}) and widths Γ (cm^{-1}) of the autoionization resonances of configuration YbI $7s6p$ (see text).

pressure. In **Table 3**, we list the data on energies (counted from the energy of the ground state Yb $4f^{14}$) excited AS with double excited valence shell: E1—RMBPT data (Ref. [30, 31]), E2—our theory.

In **Tables 4** and **5**, we list the analogous data for energies and widths of the other AS. In whole, analysis of these data shows that the results of our theory and the theory of Ivanov et al. regarding energies are in reasonable agreement with the experimental data.

However, in respect of the AS widths there is a reasonably large discrepancy between the theories, which is associated with the use of different basis of orbitals, different degree of accounting for the correlation effects, including continuum pressure (accounted channels $nln'n'$-$nl\varepsilon s(p,d,f)$).

Config.	J	E1 Theor.	E2	Exp: E_{exp}
$6p_{1/2}{}^2$	0	−1067	−1064	−1062.7
$6p_{3/2}{}^2$	2	−987	−1004	−1008.9
$6p_{1/2}6p_{3/2}$	1	−1054	−1050	−1049
$6p_{1/2}6p_{3/2}$	2	−1032	−1036	−1039.5
$5d_{3/2}{}^2$	2	−1034	−1032	−1010.76
$5d_{3/2}5d_{5/2}$	2	−994	−995	−994.63
$5d_{3/2}5d_{5/2}$	3	−1030	−1032	−1032.47

Table 3. Energies (10^2 cm^{-1}) of the YbI AS with the double excited valence shell (see text).

Conf.	J	E1	E2	Conf.	J	E1	E2
$6p_{1/2}{}^2$	0	−1067	−1064	$6p_{3/2}5d_{5/2}$	3	−963	−962
$6p_{3/2}{}^2$	0	−920	−918	$6p_{3/2}5d_{5/2}$	4	−1062	−1061
$6p_{3/2}{}^2$	2	−987	−1004	$5d_{3/2}{}^2$	0	−981	−982
$6p_{1/2}6p_{3/2}$	1	−1054	−1050	$5d_{3/2}{}^2$	2	−1034	−1032
$6p_{1/2}6p_{3/2}$	2	−1032	−1036	$5d_{5/2}{}^2$	0	−961	−963
$6p_{1/2}5d_{3/2}$	1	−1077	−1072	$5d_{5/2}{}^2$	2	−970	−968
$6p_{1/2}5d_{3/2}$	2	−1075	−1069	$5d_{5/2}{}^2$	4	−861	−859
$6p_{1/2}5d_{5/2}$	2	−1007	−1004	$5d_{3/2}5d_{5/2}$	1	−980	−982
$6p_{1/2}5d_{5/2}$	3	−1119	−1115	$5d_{3/2}5d_{5/2}$	2	−994	−995
$6p_{3/2}5d_{3/2}$	0	−1020	−1017	$5d_{3/2}5d_{5/2}$	3	−1030	−1032
$6p_{3/2}5d_{3/2}$	1	−1014	−1012	$5d_{3/2}5d_{5/2}$	4	−1024	−1026
$6p_{3/2}5d_{3/2}$	2	−914	−913	$7s_{1/2}6p_{1/2}$	0	−889	−886.4
$6p_{3/2}5d_{3/2}$	3	−1039	−1035	$7s_{1/2}6p_{1/2}$	1	−887	−886
$6p_{3/2}5d_{5/2}$	1	−949	−948	$7s_{1/2}6p_{3/2}$	1	−851	−849
$6p_{3/2}5d_{5/2}$	2	−1118	−1116	$7s_{1/2}6p_{3/2}$	2	−861	−860

Table 4. Energies (10^2 cm^{-1}) of the YbI AS with the double excited valence shell (see text).

Conf.	J	Term	Γ1	Γ2	Conf.	J	Term	Γ1	Γ2
$6p^2_{3/2}$	0	$^1S_0$	5.4	5.69	$6p_{3/2}5d_{3/2}$	2	$^1D_2{}^0$	0.20	0.52
$6p_{3/2}5d_{5/2}$	1	$^1P_1{}^0$	5.7	5.95	$5d^2_{5/2}$	0	$^1S_0$	3.30	3.63
$6p_{3/2}5d_{5/2}$	3	$^1F_3{}^0$	1.60	1.98	$5d^2_{5/2}$	2	$^3P_2$	0.40	0.73
$5d^2_{3/2}$	0	$^3P_0$	0.01	0.05	$5d^2_{5/2}$	4	$^1G_4$	0.90	1.74
$5d_{3/2}5d_{5/2}$	1	$^3P_1$	–	0.0008					

Table 5. Widths (cm^{-1}) of the YbI AS with the double excited valence shell.

The analysis shows that the state of the $5d_{3/2}5d_{5/2}$ ($J = 1$) having an abnormally small width is AS, due to the fact that its decay is prohibited in the non-relativistic limit. This remarkable fact has been found by Ivanova et al. for the first time. From the numerical viewpoint, this effect can be linked with the presence of multiple oscillations of the wave functions in the electron core area.

On the other hand, we are talking about the phenomenon, inherent in heavy atoms and associated with a complex energetics of their valence shells. This fact in principle largely explains unusual and very rich physics of the AS in the lanthanide (rare-earth) atoms. Furthermore in **Tables 6** and **7**, we list the results of computing the energies and widths of autoionization resonances $4f^{13}$ $[^2F_{7/2}]6s^2np[5/2]_2$, $4f^{13}$ $[^2F_{7/2}]6s^2nf[5/2]_2$, resulting from the excitation of electrons in $4f$-shell. For comparison, there is data from experimental measurements via three-photon laser polarization spectroscopy method (see [24, 25] and references therein).

Here, one should pay attention to the smallness of the widths of desired resonances, which has not been explained in the literature in detail. In our opinion, this is due to the complex energetics of the studied atom (heavy multi-electron core), causing some unusual physics of AS and mechanisms of their decay, especially in comparison with the usual standards of

n	E_{exp}	Γ_{exp}	E (our theory)	Γ (our theory)
12	70120.5	1.5	70,121	1.7
13	70482.0	0.4	70,483	0.5
15	70914.8	1.2	70,916	1.4
20	71428.1	0.6	71,429	0.7
25	71612.5	1.3	71,611	1.5
26	71633.3	0.6	71,631	0.8
30	71698.8	0.5	71,697	0.7
31	71710.3	0.4	71,712	0.5
34	–	–	71,741	0.3
35	–	–	71,748	0.5

Table 6. Energies and widths (cm^{-1}) of autoionization resonances $4f^{13}$ $[^2F_{7/2}]6s^2np[5/2]_2$, resulting from the excitation of electrons in $4f$-shell.

n	E_{exp}	Γ_{exp}	E (our theory)	Γ (our theory)
12	70963.6	0.5	70,965	0.7
13	71105.0	0.4	71,107	0.5
15	71312.2	1.4	71,313	1.6
20	71559.1	0.8	71,561	0.9
25	71672.5	0.5	71,673	0.8
26	71687.5	0.5	71,689	0.7
30	71732.4	0.4	71,734	0.5
31	71741.2	0.5	71,740	0.6
34	–	–	71,763	0.8
35	–	–	71,770	0.5

Table 7. Energies and widths (cm^{-1}) of autoionization resonances $4f^{13}$ $[^2F_{7/2}]6s^2nf[5/2]_2$, resulting from the excitation of electrons in 4*f*-shell.

atomic spectroscopy (typical AS widths for the He atom, atoms of the inert gases, alkali and alkali-earth atoms reach dozens and hundreds cm^{-1}).

3.2. Spectroscopy of excited and autoionization states in thulium

In contrast to the ytterbium atom, the thulium atom has been studied to a much lesser extent. Moreover, it is not difficult to understand that this atom is theoretically more complex than the ytterbium atom. In Refs. [32–41], it has been shown that two pairs of low-lying ionization limits (with vacancy states in the $4f^{14}$ shell: $4f^{-1}_{7/2}4f^{-1}_{5/2}$; here and below we write $4f_j^{-1}$ instead of $4f_j^{13}$) cause two main types of autoionization decay (**Figure 3**):

1. the classical decay channel of Beutler-Fano (BFD)

$$4f^{-1}_{5/2}6s_{1/2}(J_{12})nl \rightarrow 4f^{-1}_{7/2}6s_{1/2}(J_{12'})\mathrm{Tm}^+ + \varepsilon l_e j_{e'}\, n > 7, J_{12} = 2;3, J_{12'} = 3;4, \quad (6)$$

2. the re-orientation-type decay (ROD) channel

$$4f_j^{-1}6s_{1/2}(J_{12})nl \rightarrow 4f_j^{-1}6s_{1/2}(J_{12'})\mathrm{Tm}^+ + \varepsilon l_e j_{e'}\ n > 25,\ J_{12} = 3,\ J_{12'} = 2;4,\ \mathrm{j} = 5/2, 7/2 \quad (7)$$

Here ROD refers to re-orientation-type AS decay. The states $4f^{-1}_{5/2}6s_{1/2}(J_{12} = 3)nl$ are exposed simultaneously to BFD and ROD-decays. In contrast to BFD-decay, ROD-decay is a low energy process, preserving one-electron quantum numbers of the atomic core: $4f_j^{-1}$ and $6s_{1/2}$. The ROD-decay can be of both monopole and quadrupole nature.

This refers to the QPs interaction multi-polarity that causes the AS decay. The states with $J_{12} = 2;4$ are unable to disintegrate through ROD channel. However, their mixing with the states of the

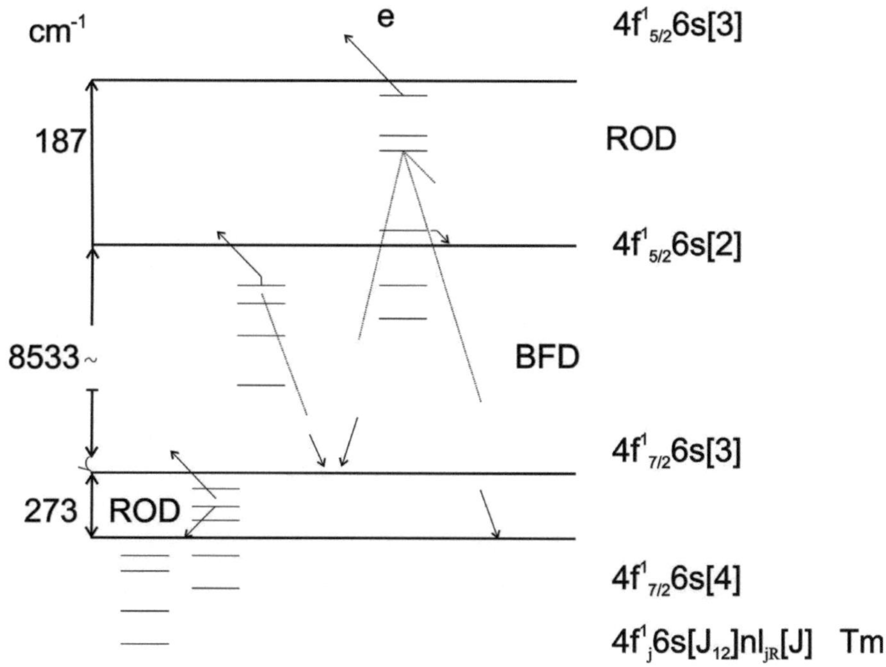

Figure 3. Position of the low-lying Tm atom $4f^{1}6snl$ ionization limits and scheme of autoionization decay of Tm atom Rydberg states $4f^{13}6snl$.

ROD decaying type can greatly improve monopole ROD decay of the latter. For these Rydberg series, there is only possible AS decay channel—re-orientation one ROD. In **Table 8**, we list the results of computing the energies E2 (10 cm^{-1}) and widths Γ2 (cm^{-1}) of the AS $4f^{-1}{}_j\ 6s(J_{12})nsnp[J]$ of the Tm atom with $n = 25$–35, namely, the AS $4f^{13}_{7/2}6s_{1/2}(3)ns_{1/2}[J]$, for which the ROD-decay is the only autoionization decay channel. For comparison, we list the calculated data obtained by Ivanov et al. (E1, Γ1) [32–41] and Glushkov et al. (E, Γ3) [6].

In **Tables 9** and **10**, we list our data on the widths and energies of the Tm AS $4f^{13}_{7/2}6s_{1/2}(3)np_j[J]$ and $4f^{13}_{5/2}6s_{1/2}(2)ns_{1/2}[J]$, for which BFD is the only autoionization decay channel. In Table 10,

	J = 5/2			J = 5/2			J = 7/2	J = 7/2
n	Γ1	Γ3	Γ2	E1	E3	E2	Γ2	E2
25	1.18 (−5)	1.29 (−5)	1.25 (−5)	4985	4981	4983	1.58 (−2)	4986
26	–	–	1.13 (−5)	–	–	4975	1.34 (−2)	4988
30	5.77 (−6)	6.72 (−6)	6.12 (−6)	4995	4993	4994	3.98 (−3)	4995
33	–	–	3.79 (−6)	–	–	4996	1.58 (−2)	4998
35	–	–	3.21 (−6)	–	–	4998	3.18 (−3)	5000

Table 8. Widths and energies of the Tm AS $4f^{13}_{7/2}6s_{1/2}(3)ns_{1/2}[J]$.

(j,J)	(3/2, 3/2)		(1/2, 5/2)		(3/2, 5/2)	
n	Г2	E2	Г2	E2	Г2	E2
25	4.68 (−5)	49,862	1.40 (−1)	49,858	1.92 (−1)	49,865
26	4.22 (−5)	49,877	1.33 (−1)	49,874	1.75 (−1)	48,979
30	2.42 (−5)	49,939	1.03 (−1)	49,937	1.07 (−1)	49,941
33	1.80 (−5)	49,971	7.54 (−2)	49,968	8.20 (−2)	49,972
35	1.39 (−5)	49,992	5.72 (−2)	49,990	6.59 (02)	49,993
(j, J)	(1/2, 7/2)		(3/2, 7/2)		(3/2, 9/2)	
n	Г2	E2	Г2	E2	Г2	E2
25	3.72 (−2)	49,848	3.46 (−1)	49,867	3.98 (−1)	49,869
26	3.45 (−2)	49,863	3.24 (−1)	49,884	3.71 (−1)	49,886
30	2.38 (−2)	49,938	2.38 (−1)	49,952	2.62 (−1)	49,953
33	2/12 (−2)	49,961	2.05 (−1)	49,977	2.26 (−1)	49,978
35	1.76 (−2)	49,982	1.56 (−1)	49,992	1.729 (−1)	49,993

Table 9. Widths and energies of the Tm AS $4f^{13}_{7/2}6s_{1/2}(3)np_j[J]$ (our data).

	J = 3/2	J = 3/2	J = 5/2	J = 5/2
n	Г2	E2	Г2	E2
25	2.64 (−5)	5836	5.32 (−5)	5838
30	1.27 (−5)	5845	2.78 (−5)	5846
35	8.89 (−6)	5850	1.54 (−5)	5852

Table 10. The widths (cm^{-1}) and energies (10^2 cm^{-1}) of the Tm AS $4f^{13}_{5/2}6s_{1/2}(2)ns_{1/2}[J]$.

we list our data on the widths and energies (in cm^{-1}) of AS $4f^{13}_{5/2}6s_{1/2}(3)ns_{1/2}[J]$, which can decay in both the ROD and the BFD channels.

4. Conclusions

We have presented a brief review of the theoretical and experimental papers on spectroscopy of the rare-earth elements of ytterbium and thulium and some results of theoretical studying their spectra. Computing energy level positions and autoionization resonance widths and energies is fulfilled on the basis of the generalized relativistic energy approach and relativistic many-body PT with the Dirac-Kohn-Sham zeroth approximation and accurate accounting for the exchange-correlation and relativistic corrections. Spectral data for some autoionization resonances, such as the Rydberg ones, or autoionization states with the double excited valence shell and others are of a great interest as they reveal a sufficiently high complexity of rare-earth elements (in particular, ytterbium and thulium) from the atomic spectroscopic viewpoint, indicating a rather complex dynamics of the decay of such states. Namely, these states play an important role in various atomic elementary processes in plasmas and gases. The Rydberg autoionization states are of considerable interest, for example, for the creation of new types of lasers, studies on the laser separation of heavy isotopes and nuclear isomers and the corresponding applications such as an analysis of trace quantities of rare-earth elements, etc. It should also be noted that the dynamics of the decay of autoionization resonances in ytterbium, thallium and other rare-earth elements in an external electromagnetic (laser) field is expected to be extremely complex and unusual. In any case, further study of the spectra, characteristics of the radiation and autoionization decay of the autoionization states in spectra of the rare-earth atoms is extremely important and actual.

Author details

Andrey A. Svinarenko*, Alexander V. Glushkov, Olga Yu. Khetselius, Valentin B. Ternovsky, Yuliya V. Dubrovskaya, Anna A. Kuznetsova and Vasily V. Buyadzhi

*Address all correspondence to: svinarenkoaa@gmail.com

Odessa State Environmental University, Odessa, Ukraine

References

[1] www.rareelementresources.com/rare-earth-elements

[2] Bourzac K. Can the US Rare-Earth Industry Rebound?. Technology Review. 2010 October

[3] Martin W. NIST Spectra Database, Version 2.0 (http://physics.nist.gov.asd). Washington: NIST; 2004

[4] Cowan RD. The Theory of Atomic Structure Spectra. Berkeley: University of California Press; 1981

[5] Grant IP. Relativistic Quantum Theory of Atoms and Molecules, Theory and Computation. Vol. 40. Berlin: Springer; 2007

[6] Glushkov AV. Relativistic Quantum Theory. Relativistic Quantum Mechanics of Atomic Systems. Odessa: Astroprint; 2008. 700 p

[7] Glushkov AV. Atom in Electromagnetic Field. Kiev: KNT; 2005. 400 p

[8] Khetselius OY. Hyperfine Structure of Atomic Spectra. Odessa: Astroprint; 2008. 208 p

[9] Kohn W, Sham S. Quantum density oscillations in an inhomogeneous electron gas. Physical Review A. 1965;**137**:1697–1710

[10] Safronova UI, Johnson W, Derevianko A. Relativistic many-body calculations of energy levels, hyperfine constants, electric-dipole matrix elements, static polarizabilities for alkali-metal atoms. Physical Review A. 1999;**60**:4476–4486

[11] Khetselius OY, Florko TA, Svinarenko AA, Tkach TB. Radiative and collisional spectroscopy of hyperfine lines of the Li-like heavy ions and Tl atom in an atmosphere of inert gases. Physica Scripta. 2013;**T153**:014037

[12] Khetselius OY. Spectroscopy of cooperative electron-gamma-nuclear processes in heavy atoms: NEET effect. Journal of Physics: Conference Series. 2012;**397**:012012

[13] Dzuba V, Flambaum V, Silvestrov P, Sushkov D. Many-body perturbation theory calculations in atoms with open shells. Physical Review A. 1991;**44**:2828–2831

[14] Camus P, Debarre A, Morillon C. Two-photon absorption spectroscopy in ytterbium. Journal of Physics B: Atomic and Molecular Physics. 1978;**11**:L395–L398

[15] Wyart J-F, Camus P. Extended analysis of the emission spectrum of neutral ytterbium (Yb I). Physica Scripta. 1979;**20**(1):43–52

[16] Spector N. Configurations 4fN–16s26p in neutral gadolinium, dysprosium, erbium, and ytterbium. Journal of the Optical Society of America A. 1971;**61**(10):1350–1354

[17] Camus P, Debarre A, Morillon C. Highly excited levels of neutral ytterbium. I. Two-photon and two-step spectroscopy of even spectra. Journal of Physics B: Atomic and Molecular Physics. 1980;**13**(6):1073–1087

[18] Aymar A, Debarre A, Robaux O. Highly excited levels of neutral ytterbium. II. Multichannel quantum defect analysis of odd- and even-parity spectra. Journal of Physics B: Atomic and Molecular Physics. 1980;**13**:1089–1096

[19] Blondel C, Champeau RJ, Delsart C. Photoionisation laser spectroscopy of ytterbium atoms in the presence of a DC electric field. Journal of Physics B: Atomic and Molecular Physics. 1985;**18**(12):2403–2414

[20] Baig MA, Connerade JP. The interchannel interaction between single excitation from 4f14 and double excitation from 6s2 in Yb I. Journal of Physics B: Atomic and Molecular Physics. 1984;**17**:L469–L474

[21] Baig MA, Ahmad S, Griesmann U, Connerade JP, Bhatti SA, Ahmad N. Inner shell and double excitation spectrum of ytterbium involving the 4f and 6s subshells. Journal of Physics B: Atomic and Molecular Physics. 1992;**25**(2):321–332

[22] Karaçoban B, Özdem L. Energies, landé factors, and lifetimes for some excited levels of neutral Ytterbium (Z = 70). Acta Physica Polonica B. 2011;**119**(3):342–353

[23] Maeda H, Matsuo Y, Takami M, Suzuki A. Optical-microwave double-resonance spectroscopy of highly excited Rydberg states of Yb. Physical Review A. 1992;**45**:1732–1742

[24] Yi J, Lee J, Kong HJ. Autoionization states of the ytterbium atom by three-photon polarization spectroscopy. Physical Review A. 1995;**51**(4):3053–3057

[25] Yi J, Park H, Lee J. Investigation of even parity autoionization states of ytterbium atoms by two-photon ionization spectroscopy. Journal of the Korean Physical Society. 2001;**39** (5):916–920

[26] Xu CB, Xu XY, Huang W, Chen DY. Rydberg and autoionization states of neutral ytterbium. Journal of Physics B: Atomic and Molecular Physics. 1994;**27**(17):3905–3913

[27] Raheel A, Yaseen M, Nadeem A, Bhatti SA, Baig MA. Two-colour three-photon excitation of the 6snf 1,3F3 and 6snp 1P1, 3P1,2 Rydberg levels of YbI. Journal of Physics B: Atomic, Molecular and Optical Physics. 1999;**32**(4):963–965

[28] Letokhov VS. Nonlinear Selective Photoprocesses in Atoms and Molecules. Moscow: Nauka; 1983. 408 p

[29] Bekov GI, Vidolova-Angelova EP, Ivanov LN, Letokhov VS, Mishin VI. Laser spectroscopy of narrow doubly excited autoionization states of ytterbium atoms. Soviet Physics—JETP. 1981;**53**(3):441–447

[30] Vidolova-Angelova EP, Ivanov LN, Letokhov VS. Application of model potential method in calculating Rydberg states of rare-earth elements Tm, Yb, Lu and their ions. Journal of Physics B: Atomic, Molecular and Optical Physics. 1982;**15**(73):981–991

[31] Vidolova-Angelova EP, Ivanov LN, Letokhov VS. Application of model potential method in calculating Rydberg states of rare-earth elements Tm, Yb, Lu and their ions. Journal of Physics B: Atomic, Molecular and Optical Physics. 1984;**17**(6):953–961

[32] Vidolova-Angelova EP, Ivanov LN, Ivanova EP, Angelov DA. Relativistic perturbation method for studying radiation decay of highly-excited many-electron atoms: Tm. Journal of Physics B: Atomic, Molecular and Optical Physics. 1986;**19**:2053–2069

[33] Vidolova-Angelova EP, Ivanov LN, Angelov DA. Autoionization decay of excited Rydberg Tm states. Journal of Physics B: Atomic, Molecular and Optical Physics. 1988;**21**:3877–3890

[34] Vidolova-Angelova EP, Ivanov LN. Autoionization Rydberg states of thulium. Re-orientation decay due to monopole interaction. Journal of Physics B: Atomic, Molecular and Optical Physics. 1991;**24**:4147–4158

[35] Ivanova EP. Energy levels of ions of silver and rhodium isoelectronic sequences with Z < 86. Optics and Spectroscopy. 2007;**103**(5):733–740

[36] Ivanova EP. Spectroscopic constants of VUV laser transitions in ions of the palladium isoelectronic sequence. Optics and Spectroscopy. 2003;**94**(2):151–156

[37] Ivanova EP, Ivanov AL. Theoretical search for optimal pump parameters for observing spontaneous radiation amplification on the lambda=41.8-nm transition of Xe IX in plasma. Quantum Electronics. 2004;**34**(11):1013–1017

[38] Ivanova EP, Ivanov LN, Aglitsky EV. Modern trends in spectroscopy of multi-charged ions. Physics Reports. 1988;**166**(N6):315–390

[39] Ivanov LN, Ivanova EP, Knight LV, Molchanov AG. Spectrum of plasma containing Ne- and Na-like ions: consistent account for Rydberg and autoionizing Rydberg series in balance equations. Physica Scripta. 1996;**53**(6):653–667

[40] Glushkov AV, Khetselius OY, Svinarenko AA. Theoretical spectroscopy of autoionization resonances in spectra of lanthanides atoms. Physica Scripta. 2013;**T153**:014029

[41] Svinarenko AA. Study of spectra for lanthanides atoms with relativistic many-body perturbation theory: Rydberg resonances. Journal of Physics: Conference Series. 2014;**548**:012039

[42] Ivanova EP, Grant IP, Rose SJ. Gain evaluation for the Ni-like tantalum X-ray laser. X-ray Lasers Proc. 1999;383-386

[43] Ivanova EP, Grant IP. Effects of oscillator strength transfer and radiative decay cancellation in Ne-like and Ni-like sequences. Application to X-ray laser modeling and plasma diagnostics. X-ray Lasers Proc. 1999;395-398

[44] Ivanova EP, Ivanov AL, Zinoviev NA, Knight LV. Proposal for X-ray laser pumped by a powerful short-pulse-drive laser. X-ray Lasers Proc. 1999;399-402

[45] Ivanova EP, Ivanov AL, Zinoviev NA, Knight LV. Proposal for the X-ray laser pumped by a powerful short-pulse drive laser. SPIE. 1999;**3735**(1):266–275

[46] Ivanova EP, Zinoviev AN. The possibility of X-ray lasers based on the innershell transitions of Ne-like ions. Physics Letters A. 2000;**274**(5):239–246

[47] Ivanova EP, Zinoviev NA. Time history of gain calculations in radiative-collisional model for X-ray lasers. Journal de Physique IV. 2001;**11**(2):151–154

[48] Ivanova EP, Zinoviev NA, Knight LV. Theoretical investigation of a Ni-like xenon X-ray laser in the 13-14 nm range. Quantum Electronics. 2001;**31**(8):683–688

[49] Ivanov LN, Ivanova EP. Extrapolation of atomic ion energies by model potential method: Na-like spectra. Atomic Data and Nuclear Data Tables. 1979;**24**(2):95–121

[50] Driker MN, Ivanova EP, Ivanov LN, Shestakov AF. Relativistic calculation of spectra of 2-2 transitions in O-and F-like atomic ions. Journal of Quantitative Spectroscopy and Radiative Transfer. 1982;**28**(6):531–535

[51] Ivanov LN, Podobedova LI. Formally exact perturbation theory with a model potential as a zeroth approximation. I. Transition energies in Fe ions including effects of inner-shell electrons. Journal of Physics B: Atomic and Molecular Physics. 1977;**10**(6):1001–1013

[52] Ivanov LN, Safronova UI, Senashenko VS, Viktorov DS. The radiationless decay of excited states of atomic systems with two K-shell vacancies. Journal of Physics B: Atomic and Molecular Physics. 1978;**11**(6):L175–L179

[53] Ivanov LN, Letokhov VS. Selective ionization of atoms in optical and electric fields. Soviet Journal of Quantum Electronics. 1975;**5**(3):329–332

[54] Ivanov LN, Letokhov VS. Spectroscopy of autoionization resonances in heavy elements atoms. Com. Mod. Phys. D.: At. Mol. Phys. 1985;**4**:169–184

[55] Ivanova EP, Ivanov LN, Glushkov AV, Kramida AE. High order corrections in the relativistic perturbation theory with the model zeroth approximation, Mg-like and Ne-like ions. Physica Scripta. 1985;**32**(4):513–522

[56] Ivanova EP, Ivanov LN, Gurchumeliya AD, Tsirekidze MA, Tsirekidze TA. Correlation effects in heavy multi-electron ions represented by 4-4 transitions in the Zn-like isoelectronic sequence. Journal of Physics B: Atomic and Molecular Physics.1985;**18**(8):1467–1482

[57] Ivanova EP, Glushkov AV. Theoretical investigation of spectra of multicharged ions of F-like and Ne-like isoelectronic sequences. Journal of Quantitative Spectroscopy & Radiative Transfer. 1986;**36**(N2):127–145

[58] Ivanova EP, Ivanov AL. A superpowerful source of far-ultraviolet monochromatic radiation. JETP. 2005;**100**(5):844–856

[59] Ivanova EP, Ivanov AL, Pakhomova TE. X-ray laser at 10-15 nm in Pd-like ions Er XXIII-Re XXX. X-ray Lasers Proc. 2007;353-359

[60] Glushkov AV, Svinarenko AA, Khetselius OY, Buyadzhi VV, Florko TA, Shakhman AN. Relativistic quantum chemistry: Advanced approach to construction of the Green's function of the Dirac equation with complex energy and mean-field nuclear potential. In: Nascimento M, Brändas E, Maruani J, Delgado-Barrio G, editors. Frontiers in Quantum Methods and Applications in Chemistry and Physics. Series: Progress in Theoretical Chemistry and Physics. Vol. 29. Berlin: Springer; 2015. pp. 197–218

[61] Glushkov AV, Malinovskaya SV, Gurnitskaya EP, Khetselius OYu, Dubrovskaya YuV. Consistent quantum theory of recoil induced excitation and ionization in atoms during capture of neutron. Journal of Physics: Conference Series. 2006;**35**:425–430

[62] Ivanov LN, Ivanova EP, Knight L. Energy approach to consistent QED theory for calculation of electron-collision strengths. Physical Review A. 1993;**48**(9):4365–4374

[63] Glushkov AV, Ivanov LN, Letokhov VS. Nuclear quantum optics. Preprint of Institute for Spectroscopy of the USSR Academy of Sciences (ISAN), Moscow-Troitsk. 1991;**AS-N4**:1–16

[64] Glushkov AV, Khetselius OY, Loboda AV, Svinarenko AA. QED approach to atoms in a laser field: Multi-photon resonances and above threshold ionization In: Wilson S, Grout PJ. Maruani J, Delgado-Barrio G, Piecuch P, editors. Frontiers in Quantum Systems in Chemistry and Physics. Series: Progress in Theoretical Chemistry and Physics. Vol. 18. Berlin: Springer; 2008. pp. 543–560

[65] Basov NG, Letokhov VS. Optical frequency standards. Soviet Physics Uspekhi. 1969;**11**(12):855–880

[66] Letokhov VS. Laser induced processes in spectroscopy, isotope separation and photochemistry. Soviet Physics Uspekhi. 1986;**29**(1):70–81

[67] Glushkov AV, Ivanov LN. A broadening of the thulium atom autoionization resonances in a weak electric field. Preprint of Institute for Spectroscopy of the USSR Academy of Sciences (ISAN), Moscow-Troitsk. 1992;**AS-N2**:1–10

[68] Glushkov AV, Ivanov LN. Shift and deformation of radiation atomic lines in the laser emission field. Multiphoton processes. Preprint of Institute for Spectroscopy of the USSR Academy of Sciences (ISAN), Moscow-Troitsk. 1992;**AS-N3**:1–12

[69] Glushkov AV, Khetselius OY, Lopatkin YM, Florko TA, Kovalenko OA, Mansarliysky VF. Collisional shift of hyperfine line for rubidium in an atmosphere of the buffer inert gas. Journal of Physics: Conference Series. 2014;**548**:012026

[70] Svinarenko AA, Khetselius OY, Buyadzhi VV, Florko TA, Zaichko PA, Ponomarenko EL. Spectroscopy of Rydberg atoms in a Black-body radiation field: Relativistic theory of excitation and ionization. Journal of Physics: Conference Series. 2014;**548**:012048

[71] Glushkov AV, Ivanov LN, Ivanova EP. Relativistic decay of excited atomic states. Generalized energy approach. In: Autoionization Phenomena in Atoms. Moscow: Moscow University Press; 1986. pp. 58–160

[72] Glushkov AV. Relativistic and correlation effects in spectra of atomic systems. Odessa: Astroprint; 2006. 400 p

[73] Glushkov AV, Ivanov LN. Radiation decay of atomic states: Atomic residue polarization and gauge noninvariant contributions. Physics Letters A. 1992;**170**(1):33–36

[74] Glushkov AV. Negative ions of inert gases. JETP Letters. 1992;**55**(2):97–100

[75] Glushkov AV, Khetselius OYu, Svinarenko AA. Relativistic theory of cooperative muon-gamma-nuclear processes: Negative muon capture and metastable nucleus discharge. In: Hoggan P, Brandas E, Maruani J, Delgado-Barrio G, Piecuch P, editors. Advances in the Theory of Quantum Systems in Chemistry and Physics. Series: Progress in Theoretical Chemistry and Physics. Vol. 22. Berlin: Springer; 2011. pp. 51–70

[76] Glushkov AV. Advanced relativistic energy approach to radiative decay processes in multielectron atoms and multicharged ions. In: Nishikawa K, Maruani J, Brandas E, Delgado-Barrio G, Piecuch P, editors. Quantum Systems in Chemistry and Physics: Progress in Methods and Applications. Series: Progress in Theoretical Chemistry and Physics. Vol. 26. Berlin: Springer; 2012. pp. 231–254

[77] Glushkov AV, Ambrosov SV, Loboda AV, Chernyakova Yu, Svinarenko AA, Khetselius OYu. QED calculation of the superheavy elements ions: Energy levels, radiative corrections and hfs for different nuclear models. Nuclear Physics A: Nuclear and Hadronic Physics. 2004;**734**:21–28

[78] Svinarenko AA, Nikola LV, Prepelitsa GP, Tkach T, Mischenko E. The Auger (autoionization) decay of excited states in spectra of multicharged ions: Relativistic theory. AIP Conference Proceedings. 2010;**1290**:94–98

[79] Glushkov AV, Ambrosov SV, Loboda AV, Gurnitskaya EP, Khetselius OY. QED calculation of heavy multicharged ions with account for correlation, radiative and nuclear effects. In: Julien J-P, Maruani J, Mayou D, Wilson S, Delgado-Barrio G, editors. Recent Advances in Theoretical Physics and Chemistry Systems. Series: Progress in Theoretical Chemistry and Physics. Vol. 15. Berlin: Springer; 2006. pp. 285–299

[80] Glushkov AV, Khetselius OYu, Gurnitskaya EP, Loboda AV, Florko TA, Sukharev DE, Lovett L. Gauge-invariant QED perturbation theory approach to calculating nuclear electric quadrupole moments, hyperfine structure constants for heavy atoms and ions. In: Wilson S, Grout PJ. Maruani J, Delgado-Barrio G, Piecuch P, editors. Frontiers in Quantum Systems in Chemistry and Physics. Series: Progress in Theoretical Chemistry and Physics. Vol. 18. Berlin: Springer; 2008. pp. 507–524

[81] Glushkov AV. Relativistic polarization potential of a many-electron atom. Soviet Physics Journal. 1990;**33**(1):1–8

[82] Glushkov AV. Correction for exchange and correlation effects in multielectron system theory. Journal of Structural Chemistry. 1990;**31**(4):529–533

[83] Glushkov AV, Khetselius OYu, Lovett L. Electron-β-nuclear spectroscopy of atoms and molecules and chemical environment effect on the β-decay parameters. In: Piecuch P, Maruani J, Delgado-Barrio G, Wilson S, editors. Advances in the Theory of Atomic and Molecular Systems Dynamics, Spectroscopy, Clusters, and Nanostructures. Series: Progress in Theoretical Chemistry and Physics. Vol. 20. Berlin: Springer; 2010. pp. 125–152

[84] Glushkov AV, Lovett L, Khetselius OYu, Gurnitskaya EP, Dubrovskaya YuV, Loboda AV. Generalized multi configuration model of decay of multipole giant resonances applied to analysis of reaction (μ-n) on the nucleus ^{40}Ca. International Journal of Modern Physics A. 2009;**24:**611–615

[85] Glushkov AV, Ivanov LN. DC strong-field stark-effect: Consistent quantum-mechanical approach. Journal of Physics B: Atomic, Molecular and Optical Physics. 1993;**26**(16): L379–L386

[86] Glushkov AV, Ivanov LN. DC Strong Field Stark effect: Consistent quantum mechanical approach. Preprint of Institute for Spectroscopy of the USSR Academy of Sciences (ISAN), Moscow-Troitsk. 1992;**AS-N1**:1–16

[87] Glushkov AV, Loboda A, Gurnitskaya E, Svinarenko A. QED theory of radiation emission and absorption lines for atoms in a strong laser field. Physica Scripta. 2009;**T135**:014022

[88] Khetselius OYu. On possibility of sensing nuclei of the rare isotopes by means of laser spectroscopy of hyperfine structure. Sensor Electronics and Microsystem Technologies. 2008;**3**:28–33

[89] Glushkov AV, Malinovskaya SV, Chernyakova YG, Svinarenko AA. Cooperative laser-electron-nuclear processes: QED calculation of electron satellites spectra for multi-charged ion in laser field. International Journal of Quantum Chemistry. 2004;**99**(5):889–893

[90] Glushkov AV, Khetselius OY, Malinovskaya SV. Spectroscopy of cooperative laser–electron nuclear effects in multiatomic molecules. Molecular Physics. 2008;**106**(9):1257–1260

[91] Glushkov AV. Operator perturbation theory for atomic systems in a strong DC electric field. In: Hotokka M, Maruani J, Brandas E, Delgado-Barrio G, editors. Advances in Quantum Methods and Applications in Chemistry, Physics, and Biology. Series: Progress in Theoretical Chemistry and Physics. Vol. 27. Berlin: Springer; 2013. pp. 161–178

[92] Glushkov AV, Khetselius OYu, Malinovskaya SV. Optics and spectroscopy of cooperative laser-electron nuclear processes in atomic and molecular systems—New trend in quantum optics. European Physical Journal Special Topics. 2008;**160**(1):195–204

[93] Glushkov AV, Malinovskaya SV. Co-operative laser nuclear processes: Border lines effects. In: Fazio G, Hanappe F, editors. New Projects and New Lines of Research in Nuclear Physics. Singapore: World Scientific; 2003. pp. 242–250

[94] Glushkov AV, Khetselius OYu, Svinarenko AA, Prepelitsa GP. Energy approach to atoms in a laser field and quantum dynamics with laser pulses of different shape. In: Duarte FJ, editor. Coherence and Ultrashort Pulsed Emission. Rijeka: InTech; 2010. pp. 159–186

Section 3

Rare Earths Thermodynamic

Chapter 5

Thermodynamic Property Study on the Complexes of Rare-Earth Elements with Amino Aids

Zhi Cheng Tan, Quan Shi, Xue Chuan Lv and
Bei Ping Liu

Additional information is available at the end of the chapter

http://dx.doi.org/10.5772/intechopen.69473

Abstract

In this chapter, the following three rare-earth complexes with amino acids, Eu(Glu)$(Im)_5(ClO_4)_3{\cdot}3HClO_4{\cdot}\ 6H_2O$, $Nd(Gly)_2Cl_3{\cdot}3H_2O$, and $La(Glu)(Im)_6(ClO_4)_3{\cdot}4HClO_4{\cdot}4H_2O$, are synthesized and characterized by element analysis, infrared (IR) spectrum, and x-ray diffraction (XRD) analysis. The thermodynamic property studies on these complexes are performed. For the first one, $Eu(Glu)(Im)_5(ClO_4)_3{\cdot}3HClO_4{\cdot}6H_2O$, the low temperature heat capacity, phase transition, and thermodynamic functions are determined by adiabatic calorimetry. For the second one, $Nd(Gly)_2Cl_3{\cdot}3H_2O$, the molar dissolution enthalpy and standard molar enthalpy of formation are determined by isoperibol solution reaction calorimetry. For the third one, $La(Glu)(Im)_6(ClO_4)_3{\cdot}4HClO_4{\cdot}4H_2O$, the microcalorimetry is used to investigate the interaction between the complex and the *Escherichia coli* DH5α to elucidate the biological effects of the complex.

Keywords: complexes of rare-earth elements, amino acids, adiabatic calorimetry (AC), solution reaction calorimetry, microcalorimetry, thermal analysis, thermodynamic properties, heat capacity, phase transition

1. Introduction

Rare-earth (RE) elements and their compounds have been extensively investigated in last several decades, because of their important significance in sciences and the growing practical applications in industries. New frontiers concerned with RE science and technology are increasingly and rapidly developed.

It is well know that China is the country richest in RE minerals all over the world. In order to sufficiently develop and effectively utilize the RE resources, the industries concerned were in urgent need of huge numbers of basic data of RE compounds. Undoubtedly, the thermodynamic property data are absolutely necessary for the study and application of rare-earth compounds. As one part of Chinese national research projects for rare-earth science and technology, in the period of 1990–2006, our Thermochemistry Laboratory in cooperation with Microcalorimetry Research Center of Osaka University well successfully completed the research project "low temperature heat capacity and thermodynamic properties of a series of rare-earth isothiocyanate hydrates, $RE(NCS)_3{\cdot}nH_2O$ (RE = La,Ce,Pr,Nd,Sm,Gd,Yb,Y)." The research results concerned were published in international journals [1–4].

Rare-earth elements have found their applications in many areas nowadays; because of their unique properties, they have been introduced into microfertilizer, pesticide, and antibacterial agent. Accompanying these applications, rare-earth elements inevitably spread into the food chain, the biological chain, and then into the bodies of human beings. This leads people to care about and study the influence and the long-term effect of rare-earth elements on themselves. Rare-earth complexes formed with amino acid were then synthesized for this purpose, because amino acid is the basic unit comprising protein and enzyme, the fundamental and functional materials in the bodies of human beings and animals. In the last 30 years, more than 200 kinds of these complexes were synthesized, and about 50 kinds among them had their own crystallograms. However, the thermodynamic property data of these complexes were seldom reported. As we all well know, only with these data we can quantitatively describe their characteristics in terms of the energetics, for example, stable forms in different temperature range, melting process, thermal anomaly, and so on. Comparisons of these data for a series of complexes comprised by the same RE (or ligands) with different ligands (or RE) may enable us to have a deeper understanding of their properties.

Rare-earth complex coordinated with amino acids possesses many unique physiological and biochemical effects. In 1975, Anghileri et al. reported for the first time that $La(Gly)_3Cl_3{\cdot}3H_2O$ was antineoplastic [5]. Since then, the rare-earth complexes coordinated with amino acids have been synthesized and studied intensively by a variety of methods [6–10]. Nowadays, the applications of rare-earth complexes with amino acids have expanded into the fields of medicine, biology, and agriculture more widely. However, the fundamental thermodynamic property data of these complexes have rarely been reported yet, especially the data of low temperature heat capacity and standard thermodynamic functions, and standard enthalpy of formation, although these data are significant for the further applications of the rare-earth complex with amino acids.

Based on previous work about thermodynamic study of rare-earth isothiocyanate hydrates, our thermochemistry laboratory in cooperation with several national academic groups has intensively contributed to this effort for thermodynamic property studies on rare-earth complexes with amino acids, and 28 academic papers concerned with this research project have been published [11–38].

In this chapter, we present the following three papers to report our recent research results and describe in detail the modern advanced experimental calorimetric techniques used for thermodynamic property studies of rare-earth complexes.

1. Thermodynamic properties of rare-earth complex with amino acid: $Eu(Glu)(Im)_5(ClO_4)_3{\cdot}3HClO_4{\cdot}6H_2O$

2. Thermodynamic study on rare-earth complex of neodymium with glycine: $Nd(Gly)_2Cl_3 \cdot 3H_2O$

3. Microcalorimetric study on biological effect of rare-earth complex with amino acid: $La(Glu)(Im)_6(ClO_4)_3 \cdot 4HClO_4 \cdot 4H_2O$

2. Thermodynamic properties of rare-earth complex with amino acid: $Eu(Glu)(Im)_5(ClO_4)_3 \cdot 3HClO_4 \cdot 6H_2O$

2.1. Experiment

2.1.1. Synthesis and characterization of the complex

The complex, $Eu(Glu)(Im)_5(ClO_4)_3 \cdot 3HClO_4 \cdot 6H_2O$, was synthesized according to the procedures mentioned below. All starting materials used for the synthesis are the analytical reagents purchased from the Beijing Chemical Reagent Co. The rare-earth oxide (Eu_2O_3) was dissolved in an excess amount of perchloric acid (1:1), and the concentration of the solution was determined by ethylene diamine tetraacetic acid (EDTA) titration analysis. Then, solid L-Glutamic acid was added to the solution in molar ratio of Eu^{3+}:Glu = 1:1. After the pH value of the reaction mixture was carefully adjusted to about 4.0 by slow addition of NaOH solution, imidazole was added in molar ratio of Eu^{3+}:Glu:Im = 1:1:5. The solution was stirred for a further 2 h in flask. The solution was transferred to the culture dish, and the culture dish was sealed using preservative film with many tiny holes stabbed using needle. The culture dish was placed in a desiccator using phosphorus pentoxide as the desiccant. The complex formed slowly in the solution and was taken out of the solution for analysis.

An elemental analysis apparatus (Model PE-2400 II, USA) was used to measure the C, H, N of the complex, and Eu was determined by EDTA titration analysis. The Cl amount in ClO_4^- of the complex was determined by the ion chromatography (Waters Alliance® 2695 System with the Waters Conductivity Detector and Waters Micromass® Quattro micro™ Mass Spectrometer). The experimental values are: Eu (11.60%) (mass fraction), C (17.794%), N (11.4093%), H (3.304%), and Cl (15.665%), which are close to the theoretical values, Eu (11.2792%), C (17.8293%), N (11.4358%), H (3.2917%), and Cl (15.789%). The sample formula was determined to be $Eu(Glu)(Im)_5 (ClO_4)_3 \cdot 3HClO_4 \cdot 6H_2O$.

Infrared spectra of the complex and L-Glutamic acid were obtained through KBr pellets at room temperature using a Bruker Tensor 27-IR spectrophotometer. The infrared spectrum of the complex was shown in **Figure 1**. Compared with the IR spectrum of L-Glutamic acid, the νs (carboxyl) band of the complex shifted from 1410 cm^{-1} to higher wave numbers (1431 cm^{-1}), which shows that the carboxyl groups of the ligand have been coordinated to the metal ion [39]. The special absorptions of $-NH_2$ shifted from 3050 to 3240 cm^{-1} (ν -NH), from 2750 to 2740 cm^{-1} (ν -NH), and from 1560 to 1562 cm^{-1} (δNH_2^+), because a hydrogen bond formed in the complex. The spectrum also shows the wide peak symmetrical resonance frequencies, νs (N-H), shifted from 3286 to 3425 cm^{-1} down to 3166 to 3098 cm^{-1}, which is evidence of the

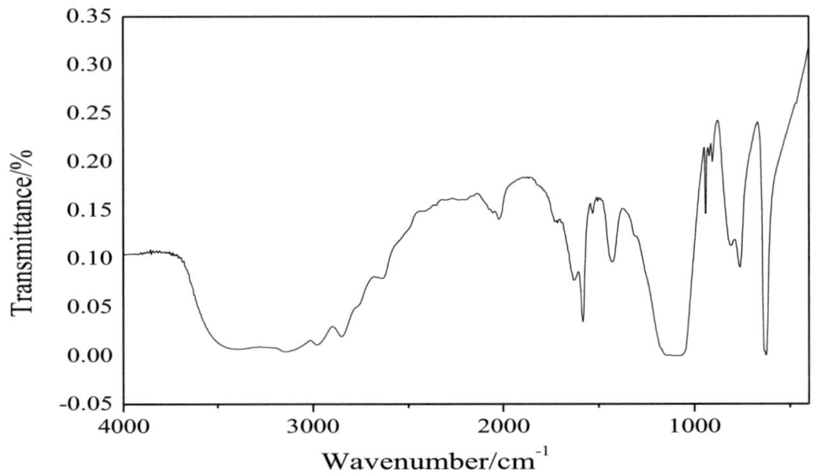

Figure 1. Infrared spectrum of the complex.

coordination of imidazole molecules [40]. A broad absorption band for ν (hydroxyl) appearing at 3400 cm^{-1} shows the presence of water molecules in the complex.

Powder X-ray diffraction (XRD) patterns were recorded using Rigaku D/Max 3400 X with Cu Kα radiation at 0.02 steps per second in angle range 2θ = 5–90, and the results were shown in **Figure 2**. Judging from the shape and number of the peak in the patterns, the sample was a kind of complex with single phase.

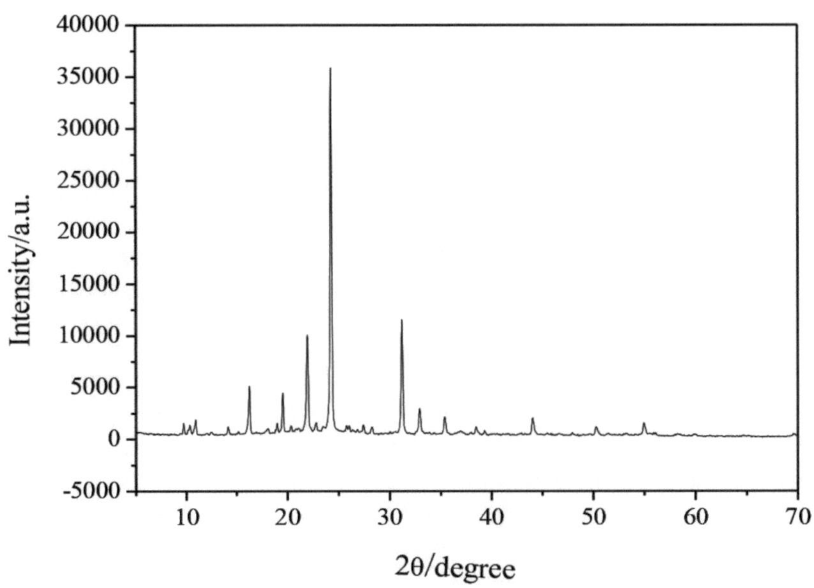

Figure 2. Powder X-ray diffraction patterns of complex.

http://dx.doi.org/10.5772/intechopen.69473

2.1.2. Adiabatic calorimetry

Adiabatic calorimetry is the most accurate approach to obtain the heat capacity data. In the present study, heat capacity measurements were carried out by a high-precision automatic adiabatic calorimeter over the temperature range of 80–350 K. The adiabatic calorimeter was established by Thermochemistry Laboratory of Dalian Institute of Chemical Physics, Chinese Academy of Sciences in PR China. The structure and principle of the adiabatic calorimeter have been described in detail elsewhere [41]. The schematic diagram of the adiabatic calorimeter is shown in **Figure 3**. Briefly, the automatic adiabatic calorimeter is mainly composed of a sample cell, a miniature platinum resistance thermometer, an electric heater, the inner and outer adiabatic shields, two sets of six-junction chromel-constantan thermopiles installed between the calorimetric cell and the inner shield and between the inner and the outer shields, respectively, and a high vacuum can. The working temperature range is 78–400 K and, if necessary, it can be cooled by liquid nitrogen. The heat capacity measurements were conducted by the standard procedure of intermittently heating the sample and alternately measuring the temperature. The heating rate and the temperature increments of the experimental points were generally controlled at 0.1–0.4 $K{\cdot}min^{-1}$ and at 1–4 K, respectively, during the whole experimental process. The heating duration was 10 min, and the temperature drift rates of the sample cell measured in an equilibrium period were kept within 10^{-3}–10^{-4} $K{\cdot}min^{-1}$ during the acquisition of heat capacity data. In order to verify the reliability of the adiabatic calorimeter, the molar heat capacities $C_{p,m}$ of the Standard Reference Material (SRM-720) (α-Al_2O_3) were measured in the range of 78–400 K. The deviation of our calibration data from those of NIST [42] was within ±0.1% (standard uncertainty). In the present study, the mass of the complex, $Eu(Glu)(Im)_5(ClO_4)_3{\cdot}3HClO_4{\cdot}6H_2O$, used for heat capacity measurement was 2.0352 g.

The result of heat capacity measurement of the complex was listed in **Table 1** and shown in **Figure 4**.

2.1.3. Other thermal analysis

A differential scanning calorimeter (DSC−141, SETARAM, France) was used to perform the thermal analysis of $Eu(Glu)(Im)_5(ClO_4)_3{\cdot}3HClO_4{\cdot}6H_2O$. In the range 100–300 K, liquid nitrogen was used as cryogen. In the range 300–700 K, nitrogen gas was taken as protective gas with flow rate of 50 $cm^3{\cdot}min^{-1}$, at the heating rate of 10 $K{\cdot}min^{-1}$. The masses of the samples used in the above two experiments were 4.03 and 4.26 mg, respectively.

Two aluminum crucibles were used and the reference crucible was empty. The calibrations for the temperature and heat flux of the calorimeter were performed prior to the experimental measurements of the sample. The temperature scale was calibrated by measuring the melting points of Hg, In, Sn, Pb, and Zn, at different heating rates, and the heat flux was calibrated by the Joule effect. Measurements of the melting temperature and the enthalpy of fusion of benzoic acid (NIST, Standard Reference Material 39i) were performed in the laboratory to check the accuracy of the instrument.

The result of DSC measurement was shown in **Figure 5**.

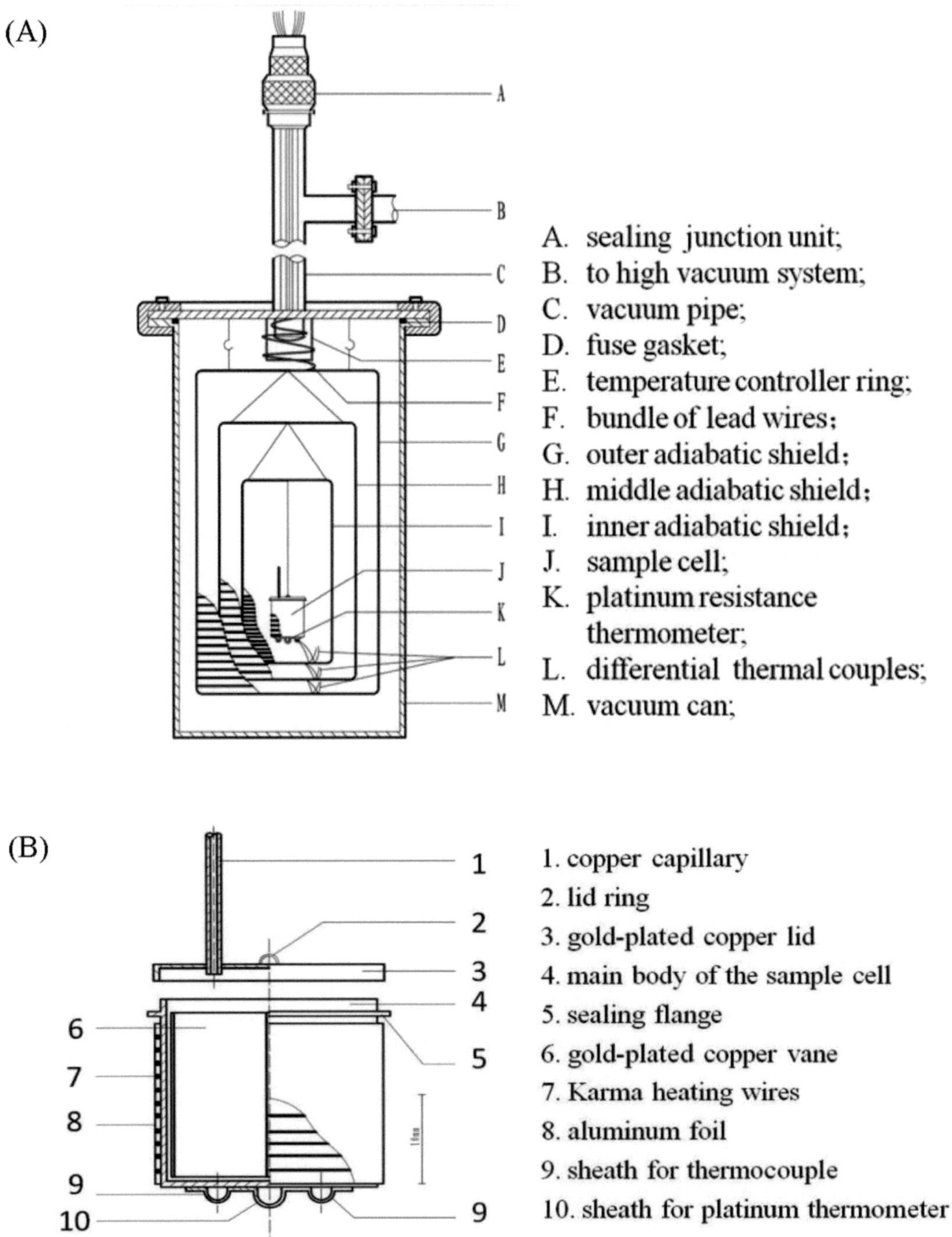

Figure 3. (A) Schematic diagram of main body of the adiabatic calorimeter. (B) Schematic diagram of sample cell of the adiabatic calorimeter.

http://dx.doi.org/10.5772/intechopen.69473

T/K	$C_{p,m}$/kJ^{-1}·mol^{-1}	*T*/K	$C_{p,m}$/kJ^{-1}·mol^{-1}
80.809	919.23	216.99	2245.5
85.210	937.21	220.72	2147.5
89.334	959.77	224.63	2043.8
92.891	985.26	228.55	1900.4
96.695	999.58	232.60	2160.4
100.86	1023.5	236.58	2483.0
103.43	1041.7	240.60	2787.8
107.12	1059.7	244.18	4772.0
112.49	1090.3	246.08	11808.8
116.85	1120.2	252.57	2262.7
120.68	1140.2	258.60	2282.7
124.34	1163.4	262.37	2297.0
128.48	1187.6	266.06	2302.0
132.40	1209.4	269.69	2305.4
136.47	1242.6	273.60	2316.5
140.42	1263.1	277.58	2330.1
144.39	1284.7	281.39	2336.3
148.49	1308.0	285.46	2355.6
152.58	1334.7	289.26	2377.3
156.48	1359.5	293.34	2416.1
160.77	1383.9	297.43	2444.2
164.78	1412.6	301.37	2490.8
168.80	1439.5	305.18	2526.2
172.87	1465.4	309.23	2571.8
176.80	1488.0	313.12	2611.6
180.71	1512.7	316.85	2661.0
184.90	1542.9	319.85	2721.8
188.89	1577.1	324.82	2779.0
192.76	1610.9	328.77	2842.9
196.79	1644.1	332.78	2906.9
200.75	1681.2	336.67	2974.1
204.67	1728.7	340.52	3051.9
208.78	1788.1	344.52	3126.9
212.99	2025.9	348.77	3209.9

Table 1. Experimental molar heat capacity of the complex from 80 to 350 K at pressure p = 0.1 MPa.

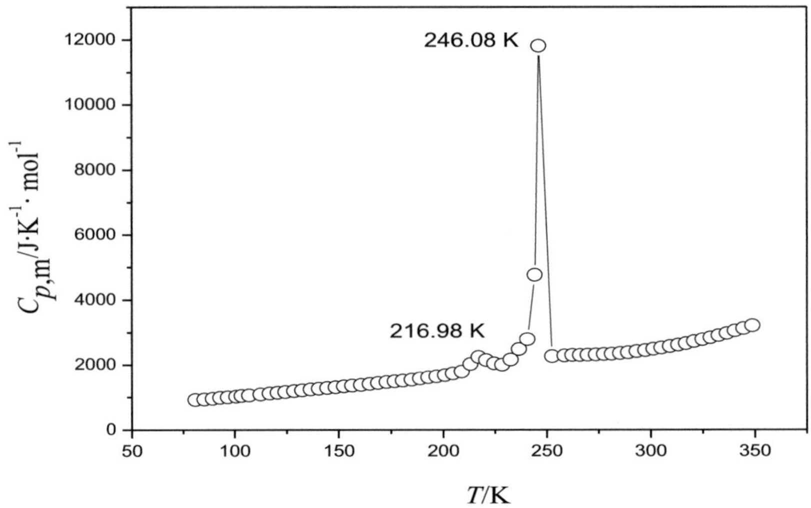

Figure 4. Experimental molar heat capacities as a function of temperature for the complex from 80 to 350 K.

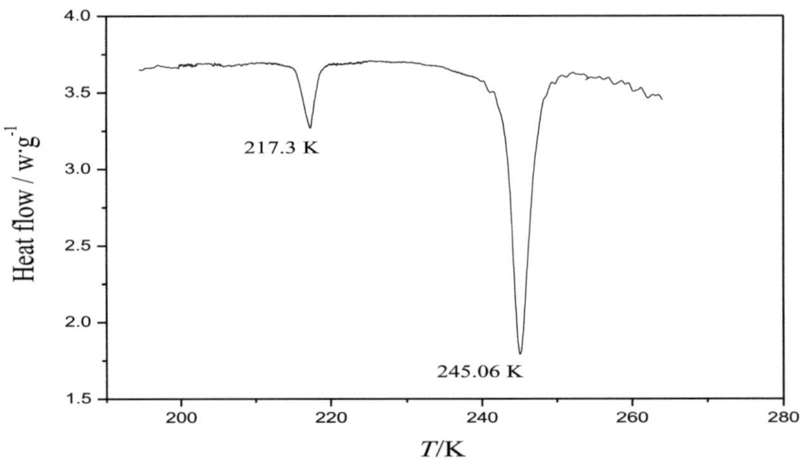

Figure 5. DSC curve of the complex from 200 to 270 K.

The thermogravimetric (TG) measurement of the complex was carried out by a TG analyzer (Model: setsys 16/18 from Setaram Company, France) at the heating rate of 10 K·min^{-1} under nitrogen atmosphere (99.999%), with flow rate of 65 cm^{3}·min^{-1}. The mass of the sample used in the experiment was 6.3 mg. Two Al_2O_3 crucibles with capacity of 100 μL were used for sample and empty cell, respectively. The calibration of TG analyzer was performed by the SRM in thermal analysis, $CaC_2O_4 \cdot H_2O(s)$.

The result of TG measurement of the complex was shown in **Figures 6** and **7**, respectively.

http://dx.doi.org/10.5772/intechopen.69473

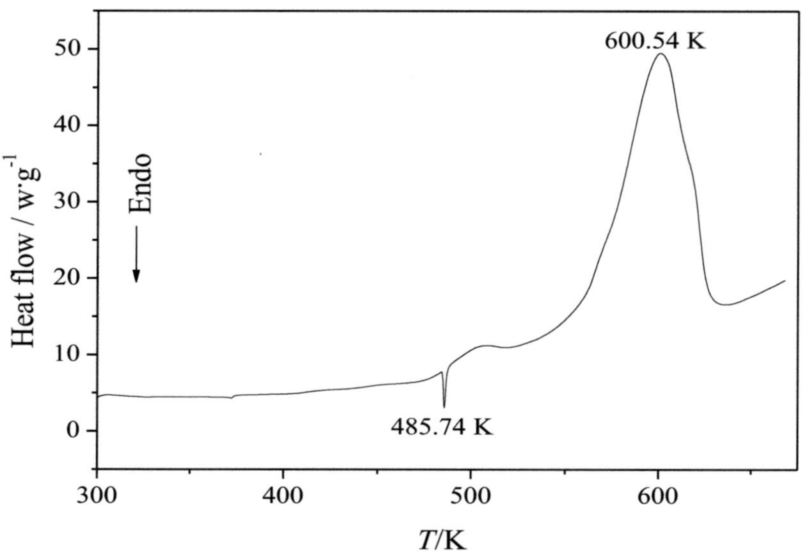

Figure 6. DSC curve of the complex from 300 to 700 K.

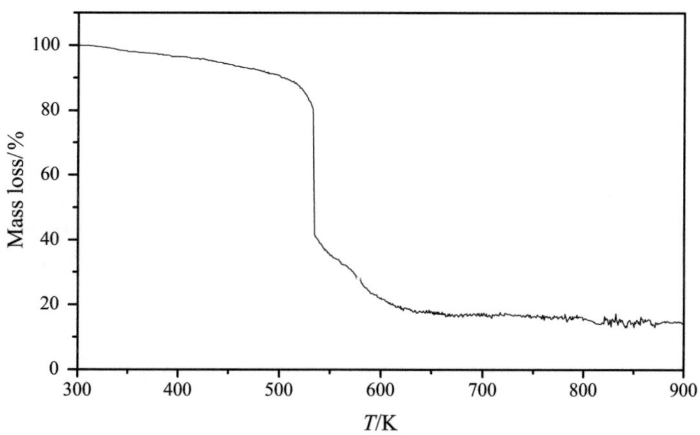

Figure 7. TG curve of the complex from 300 to 900 K.

2.2. Results and discussion

2.2.1. Phase transitions

From the view of phase temperature of the process, Yukawa et al. [43] found a phase transition of the first order and a glass transition of crystalline $(TMA)_2$ $[Sm\{Ni(Pro)_2\}_6](ClO_4)_4$ at T = (190 ± 1) K and T = (162 ± 2) K. They contributed the phase transition to orientational order/disorder process of perchlorate ions ClO_4^- and the glass transition to a freezing-in phenomenon of the reorientational motion of perchlorate ions ClO_4^-. They discussed that the cooperative interaction between the orientations of the ClO_4^- ions operates through the orientational and positional shifts of TMA ions, and thus the lattice deformation in the relevant region, associated with the orientational change of the ClO_4^- ions. The position of the ClO_4^- ions itself would shift to form preferable ionic interaction.

It can be seen from **Figure 4** that there are two thermal events in the range 80–350 K on the $C_{p,m}$-T curve. The heat capacity of the sample increases with increasing temperature in a smooth and continuous manner in the temperature range from 80 to 208 K. A little peak occurs at T = 216.98 K on the $C_{p,m}$-T curve, which is consistent with the peak (T_{trs} = 217.30 K) on the DSC curve (**Figure 5**). Such thermal event is deduced to be the phase transition of the complex. The second peak is at 246.08 K, which is also deduced to be the phase transition consistent with the peak (T_{trs} = 245.06 K) on the DSC curve.

It can be proved from **Figure 4** that there are two endothermic processes in the temperature range 100–300 K. One is at 217.30 K, and the other is at 245.06 K, which confirms the results of the heat capacity measurements (216.98 and 246.08 K). The temperature ranges and peak values of the endothermic processes correspond to those of the phase transitions in the $C_{p,m}$-T curve obtained from the heat capacity measurements. It is also verified that there are two phase transitions in the temperature range 100–350 K.

From the view of energy accompanying the process, Yukawa et al. [43] estimated the entropy of the phase transition was to be from 20 to 45 $J{\cdot}K^{-1}{\cdot}mol^{-1}$ in the range. The present work calculated that the molar entropy $\Delta_{trans}S_m$ of the first-phase transition is 27.86 ± 0.08 $J{\cdot}mol^{-1}{\cdot}K^{-1}$, which was in the range of entropy of the phase transition estimated by Yukawa et al.

The combined consideration based on the results of AC (adiabatic calorimetric) and DSC (differential scanning calorimetric) measurements, the values of thermodynamic functions, and the report of former documents leads to a conclusion that the phase transition was interpreted as a freezing-in phenomenon of the reorientational motion of perchlorate ions ClO_4^-. The endothermic process at T = 246.86 K was deduced to be the orientational order/disorder process of ClO_4^- ion [43–46].

2.2.2. Molar heat capacity, molar enthalpies, and entropies

Experimental molar heat capacities, $C_{p,m}$ against temperature T in the range 80–350 K, were listed in **Table 1** and plotted in **Figure 4**. It can be seen from the figure that no phase change or thermal decomposition was found before 206 K and after 258 K.

None-phase transitions process:

Before the solid-solid phase transitions process (from 80 to 208 K):

$$H_T - H_{298.15} = \int_{298.15}^{T} C_{p,m}(s)dT \tag{1}$$

$$S_T - S_{298.15} = \int_{298.15}^{T} [C_{p,m}(s)/T]dT \tag{2}$$

After the solid-solid phase transitions process (from 258 to 350 K):

$$H_{\mathrm{T}} - H_{298.15} = \int_{298.15}^{\mathrm{T_i}} C_{\mathrm{p,m}}(s)dT + \Delta_{trans}H_m + \int_{\mathrm{T}_f}^{\mathrm{T}} C_{\mathrm{p,m}}(s)dT \tag{3}$$

$$S_{\mathrm{T}} - S_{298.15} = \int_{298.15}^{\mathrm{T_i}} [C_{\mathrm{p,m}}(s)/T]dT + \Delta_{trans}S_m + \int_{\mathrm{T}_f}^{\mathrm{T}} [C_{\mathrm{p,m}}(s)/T]dT \tag{4}$$

where $C_{\mathrm{p,m}}$(s) is the molar heat capacity of the solid state, T_{i} is the temperature at which the solid-solid phase transition started, and T_{f} is the temperature at which the solid-solid phase transition ended.

The values of the experimental heat capacities except the phase change regions were fitted. One polynomial Eq. (5) was obtained by least squares fitting using the experimental molar heat capacities ($C_{p,\mathrm{m}}$) and the reduced temperatures (x).

$$C_{p,\mathrm{m}}/(\mathrm{J \cdot K \cdot mol^{-1}}) = 1838 + 1240x - 136.0x^2 - 806.7x^3 + 376.3\,x^4 + 765.5\,x^5 \tag{5}$$

where $x = [(T/K) - 215]/135$, x is the reduced temperature, and T is the experimental temperature. Correlation coefficient of the least squares fitting, R^2, is 0.9989 (T_i = 80 K and T_{f} =350 K).

Phase transitions process (from 206 to 258 K):

In the solid-solid phase transitions process (from 206 to 258 K), the temperature T_{trans}, molar enthalpies $\Delta_{\mathrm{trans}}H_{\mathrm{m}}$, and molar entropies $\Delta_{\mathrm{trans}}S_{\mathrm{m}}$ of the first two-phase transitions are determined according to Eqs. (6)–(10) and listed in **Table 3**.

$$\Delta_{\mathrm{trans}}H_{\mathrm{m}} = \sum_{k=1}^{f}[C_{\mathrm{p,m}}(\mathrm{trans})(T_{k+1} - T_k)] \tag{6}$$

$$\Delta_{\mathrm{trans}}S_{\mathrm{m}} = \sum_{k=1}^{f}\left[\frac{C_{\mathrm{p,m}}(\mathrm{trans})}{T_{\mathrm{trans}}}(T_{k+1} - T_k)\right] \tag{7}$$

where $C_{\mathrm{p,m}}$(trans) is the molar heat capacity during the solid-solid phase transitions process:

$$C_{\mathrm{p,m}}(\mathrm{trans}) = [C_{p,m(k)} + C_{p,m(k+1)}]/2 \tag{8}$$

T_{trans} can be seen as follows:

$$T_{\mathrm{trans}} = [(T_k + T_{k+1}]/2 \tag{9}$$

From Eqs. (6) to (9), molar enthalpies, $\Delta_{\mathrm{trans}}H_{\mathrm{m}}$, and molar entropies, $\Delta_{\mathrm{trans}}S_{\mathrm{m}}$, of the two solid-solid phase transition processes from 208 to 228 K and from 228 to 252 K were calculated as follows:

$$\Delta_{\mathrm{trans}}H_{\mathrm{m}} = \sum_{k=1}^{f}[C_{\mathrm{p,m}}(\mathrm{trans},1)(T_{k+1} - T_k)] \tag{10}$$

$$\Delta_{\text{trans}}S_{\text{m}} = \sum_{k=1}^{f}\left[\frac{C_{\text{p,m}}(\text{trans},1)}{T_{\text{trans}}}(T_{k+1}-T_k)\right] \tag{11}$$

$$C_{\text{p,m}}(\text{trans},1) = C_{\text{p,m}}(\text{trans}) - C_{\text{p,m}}(\text{trans},2) \tag{12}$$

where $C_{p,m}$(trans, 1) is the molar heat capacity caused by solid-solid phase transitions. $C_{p,m}$(trans) is the total molar heat capacity. $C_{p,m}$ (trans, 2) is the molar heat capacity induced by temperature change of substance, and its polynomial equation was obtained by least squares analysis using the experimental molar heat capacities in the non-phase transition regions from 80 to 350 K as in Eq. (5).

The molar enthalpy of the first-phase transition was calculated from 208 to 228 K as follows:

$$\Delta_{\text{trans}}H_{\text{m},1} = \sum_{k=1}^{f}[C_{\text{p,m}}(\text{trans},1)(T_{k+1}-T_k)] = (3.896 \pm 0.01)\ \text{kJ} \cdot \text{mol}^{-1} \tag{13}$$

$$\Delta_{\text{trans}}S_{\text{m},1} = \sum_{k=1}^{f}\left[\frac{C_{\text{p,m}}(\text{trans},1)}{T_{\text{trans}}}(T_{k+1}-T_k)\right] = (17.872 \pm 0.050)\text{J} \cdot \text{mol}^{-1} \cdot \text{K}^{-1} \tag{14}$$

(Standard uncertainty), where T_1 = 208.86 K and T_f = 228.55 K.

The molar enthalpy of the second-phase transition was calculated from 228 to 252 K, at 0.1 MPa as follows:

$$\Delta_{\text{trans}}H_{\text{m},2} = \sum_{k=1}^{f}[C_{\text{p,m}}(\text{trans},1)(T_{k+1}-T_k)] = (53.763 \pm 0.160)\ \text{kJ} \cdot \text{mol}^{-1} \tag{15}$$

(Standard uncertainty), where T_1 = 228.55 K and T_f = 252.57 K. The thermodynamic functions relative to the reference temperature 298.15 K, (H_T-$H_{298.15}$), and (S_T-$S_{298.15}$) were calculated in the temperature range from 80 to 350 K with an interval of 5 K at pressure 0.1 MPa, and listed in **Table 2**, using the above thermodynamic polynomial Eqs. (1)–(12).

2.2.3. *Thermal decomposition properties*

It can be seen from **Figure 6** that there are two peaks on the DSC curve from 300 to 700 K. One is a little endothermic peak (T_{trans} = 485.57 K), which was caused by decomposition of amino, and the other is a wide exothermic peak from 530 to 600 K, which was contributed to decomposition of the complex. It was consistent with the results of TG analysis.

The result of TG analysis was showed in **Figure 7**. It can be seen from the TG curve that mass loss of $Eu(Glu)(Im)_5(ClO_4)_3{\cdot}3HClO_4{\cdot}6H_2O$ began at about 298 K and ended at about 900 K. The whole process was divided into three stages. The mechanism of the decomposition was deduced as following:

$$Eu(Glu)(Im)_5(ClO_4)_3 \cdot 3HClO_4 \cdot 6H_2O \xrightarrow[8.10\%(8.0\%)]{300\sim485.88K} Eu(Glu)(Im)_5(ClO_4)_3 \cdot 3HClO_4 \quad (16)$$

$$Eu(Glu)(Im)_5(ClO_4)_3 \cdot 3HClO_4 \xrightarrow[40.90\%(39.43\%)]{531\sim536K} Eu(ClO_4)_3 \cdot 3HClO_4 \quad (17)$$

$$Eu(ClO_4)_3 \cdot 3HClO_4 \xrightarrow[25.7\%(25.45\%)]{536\sim882K} EuO_{3/2}[14.59\%(13.11\%)] \quad (18)$$

The value in the parentheses is theoretical calculation result.

T/K	$C_{p,m}$/J·mol^{-1}·K^{-1}	(H_T-$H_{298.15}$)/kJ·mol^{-1}	(S_T-$S_{298.15}$)/J·mol^{-1}·K^{-1}
80	879.70	−417.61	−2182.7
85	928.49	−413.09	−2127.0
90	970.11	−408.34	−2072.6
95	1006.0	−403.39	−2019.4
100	1037.4	−398.28	−1967.5
105	1065.6	−393.02	−1916.6
110	1091.5	−387.63	−1866.9
115	1116.0	−382.10	−1818.1
120	1139.9	−376.46	−1770.3
125	1163.8	−370.70	−1723.4
130	1188.3	−364.82	−1677.2
135	1213.9	−358.81	−1631.8
140	1240.8	−352.67	−1587.0
145	1269.4	−346.40	−1542.7
150	1299.9	−339.97	−1498.9
155	1332.3	−333.39	−1455.5
160	1366.7	−326.64	−1412.4
165	1403.1	−319.72	−1369.6
170	1441.4	−312.61	−1327.0
175	1481.5	−305.30	−1284.5
180	1523.2	−297.79	−1242.1
185	1566.3	−290.07	−1199.7
190	1610.6	−282.12	−1157.3
195	1655.9	−273.96	−1114.9
200	1701.8	−265.56	−1072.4
205	1748.2	−256.94	−1029.9
210–250	Phase transitions	\	\
255	2180.7	−100.65	−365.32

T/K	$C_{p,m}$/J·mol^{-1}·K^{-1}	$(H_T-H_{298.15})$/kJ·mol^{-1}	$(S_T-S_{298.15})$/J·mol^{-1}·K^{-1}
260	2217.6	−89.657	−322.47
265	2253.2	−78.479	−279.73
270	2287.9	−67.124	−237.11
275	2321.9	−55.598	−194.63
280	2355.6	−43.903	−152.32
285	2389.5	−32.038	−110.16
290	2424.1	−20.002	−68.162
295	2460.2	−7.7892	−26.307
298.15	2484.1	0.0000	0.0000
300	2498.6	4.6103	15.432
305	2540.2	17.210	57.097
310	2586.0	30.028	98.746
315	2637.2	43.088	140.46
320	2695.2	56.422	182.33
325	2761.5	70.067	224.49
330	2837.6	84.069	267.09
335	2925.4	98.481	310.29
340	3027.0	113.37	354.32
345	3144.3	128.80	399.42
350	3279.7	144.87	445.87

Table 2. Thermodynamic functions of the complex from 80 to 350 K at pressure p = 0.1 MPa.

Transition	T_{trans}/K	$\Delta_{trans} H_m$/kJ·mol^{-1}	$\Delta_{trans} S_m$/J·mol^{-1}·K^{-1}
Phase transition I	216.98 ± 0.65	3.896 ± 0.022	17.87 ± 0.05
Phase transition II	246.08 ± 0.74	53.76 ± 0.16	217.9 ± 0.7

Table 3. Temperature, enthalpy, and entropy of the phase transitions of the complex obtained by the heat capacity measurements from 80 to 350 K.

3. Thermodynamic study on rare earth complex of neodymium with glycine: $Nd(Gly)_2Cl_3{\cdot}3H_2O$

3.1. Experiment

3.1.1. Sample preparation and characterization

The complex, $Nd(Gly)_2Cl_3{\cdot}3H_2O$, was synthesized according to the following procedure. Firstly, rare-earth oxide (Nd_2O_3, purity 99.18%) was dissolved in chloride acid to obtain the aqueous solution of the rare-earth chloride. Following that, the aqueous solution was mixed with glycine in the molar ratio of 1:2 at about pH = 2.5 for 7–8 h. After that, the mixed solution was concentrated in a thermostat bath at 323 K until most of the water was evaporated, and then the concentrated solution was dried in a vacuum desiccator over phosphorus pentoxide for several weeks until lavender crystals appeared.

The actual amount of C, H, and N in the as-prepared sample was confirmed by elemental analysis instrument (model PE-2400 II, USA). The contents of Cl and Nd were determined by the EDTA titration and Mohr method, respectively. The experimental values are: C(10.66%), N (2.97%), H(2.49%), Cl(23.13%) and Nd(31.65%), which are close to the theoretical values of C (10.56%), N(3.08%), H(2.42%), Cl(23.39%) and Nd(31.72%). The sample formula was determined to be $Nd(Gly)_2Cl_3{\cdot}3H_2O$, and the purity obtained from the EDTA titration under the same conditions was found to be 99.78%, which was high enough to meet the requirements of the present calorimetric study. Infrared spectra of the complex $Nd(Gly)_2Cl_3.3H_2O$, $Pr(Gly)_2Cl_3{\cdot}3H_2O$, $Er(Gly)_2Cl_3{\cdot}3H_2O$ and glycine were obtained from KBr pellets at room temperature using an IR spectrophotometer (model AVATAR 370 NICOLET USA).

Infrared spectra of $Nd(Gly)_2Cl_3{\cdot}3H_2O$, $Pr(Gly)_2Cl_3{\cdot}3H_2O$, $Er(Gly)_2Cl_3{\cdot}3H_2O$, and glycine are demonstrated in **Figure 8**. Compared with the IR spectrum of glycine, the νs(carboxyl) band of $Nd(Gly)_2Cl_3{\cdot}3H_2O$ shifted from 1394 cm^{-1} to higher wave numbers (1418 cm^{-1}), which reveals that the carboxyl groups of the ligand have been coordinated to the metal ion. The special absorption of NH_2 shifted from 3105^{-1} to 3420 cm^{-1} (νN-H), from 2604 to 2580 cm^{-1} (νN-H), and from 1500 to 1477 cm^{-1} (υ_{NH_2}) indicates a hydrogen bond formed in the complex. A broad absorption band for νs(hydroxyl) appearing at 3400 cm^{-1} suggests the presence of water molecules in the complex. Meanwhile, it can be observed that main diffraction peaks of Nd

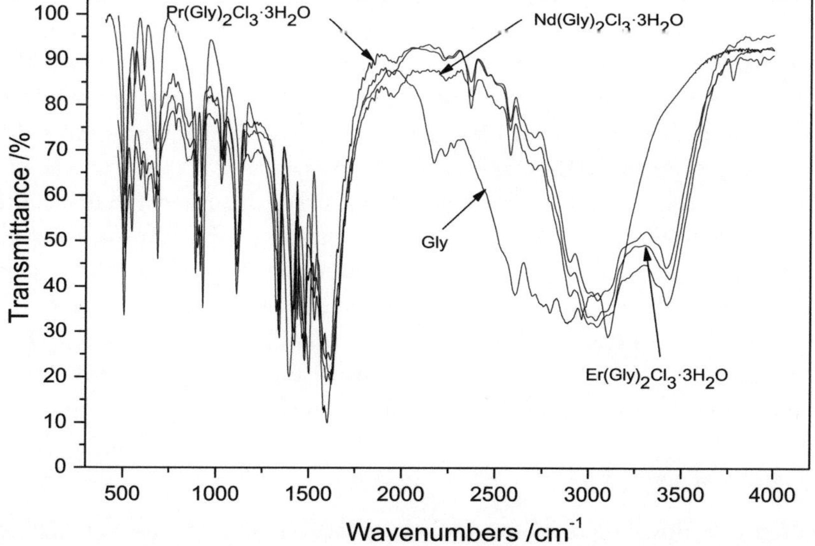

Figure 8. Infrared spectra of the complexes $Nd(Gly)_2Cl_3{\cdot}3H_2O$, $Pr(Gly)_2Cl_3{\cdot}3H_2O$, and $Er(Gly)_2Cl_3{\cdot}3H_2O$.

$(Gly)_2Cl_3{\cdot}3H_2O$ are in good agreement with that of $Pr(Gly)_2Cl_3{\cdot}3H_2O$ and $Er(Gly)_2Cl_3{\cdot}3H_2O$. The above results suggest that the structure of as-synthesized complex is similar to that of $Pr(Gly)_2Cl_3{\cdot}3H_2O$ and $Er(Gly)_2Cl_3{\cdot}3H_2O$.

3.1.2. DSC and TG measurements

Thermal analysis of $Nd(Gly)_2Cl_3{\cdot}3H_2O$ was performed using a differential scanning calorimeter (DSC-141, SETARAM, France) and a thermogravimetric analyzer (model DT-20B, Shimadzu, Japan). The DSC measurement was carried out with a heating rate of 10 K^{-1} min^{-1} under high-purity nitrogen with a flow rate of 50 mL^{-1} min^{-1}. The mass of the sample used in the experiment was 3.36 mg. The calibrations for the temperature and heat flux of the calorimeter were performed prior to the experiment. The temperature scale was calibrated by measuring the melting points of Hg, In, Sn, Pb, and Zn at different heating rates, and the heat flux was calibrated using the Joule effect. Measurement of the melting temperature and the enthalpy of fusion of benzoic acid (NIST, SRM 39i) were carried out in this laboratory to check the accuracy of the instrument. The TG measurement of the complex was conducted at a heating rate of 10 K^{-1} min^{-1} under high-purity nitrogen with a flow rate of 120 $mL{\cdot}min^{-1.}$ The mass of the sample used in the experiment was 9.47 mg. The reference crucible was filled with α-Al_2O_3. The TG-DTG equipment was calibrated by the thermal analysis of the SRM, $CaC_2O_4{\cdot}H_2O(s)$.

3.1.3. Dissolution enthalpy measurements

3.1.3.1. Isoperibol solution reaction calorimeter

The Isoperibol solution reaction calorimeter (SRC-100) used for this study was constructed in our thermochemistry laboratory and has been used for measuring enthalpies of solid-liquid, liquid-liquid dissolution reactions enthalpies of the reactant, and the product to obtain the formation enthalpies of the complex. The principle and structure of the instrument were described elsewhere [47]. The schematic diagram of the calorimeter is shown in **Figure 9**. The calorimeter mainly consisted of a water thermostat, a pyrex-glass Dewar with volume of 100 mL, a glass sample cell with volume of 2 mL, a heater for calibration and equilibration purposes, a glass-sheathed thermistor probe, an amplifier, a circuit used as an A/D converter, and a personal computer for data acquisition and processing. The Dewar vessel, equipped with a twin-blade stirrer, served as a mixing chamber, was submerged in the water thermostat. The precisions of controlling and measuring the temperature of the calorimeter are ±0.001 and ±0.0001 K, respectively. The measurements of solution-reaction enthalpy of samples were conducted under atmospheric pressure. The calorimeter was tested by measuring the dissolution enthalpies of the KCl (calorimetric primary standard) in water at 298.15 K. The mean dissolution enthalpies obtained from five experiments are 17515 ± 12 $J{\cdot}mol^{-1}$, which are in conformity with the reported data (17536 ± 9 $J{\cdot}mol^{-1}$) [48].

3.1.3.2. Dissolution and formation enthalpy measurements

The dissolution enthalpies of reactants and the products, and standard molar enthalpy of formation of $Nd(Gly)_2Cl_3{\cdot}3H_2O$ were determined through the proposed Hess thermochemical cycle, which is demonstrated in **Figure 10**. The dissolution enthalpy of H_2O (l), $\Delta_d H$, as one of

http://dx.doi.org/10.5772/intechopen.69473

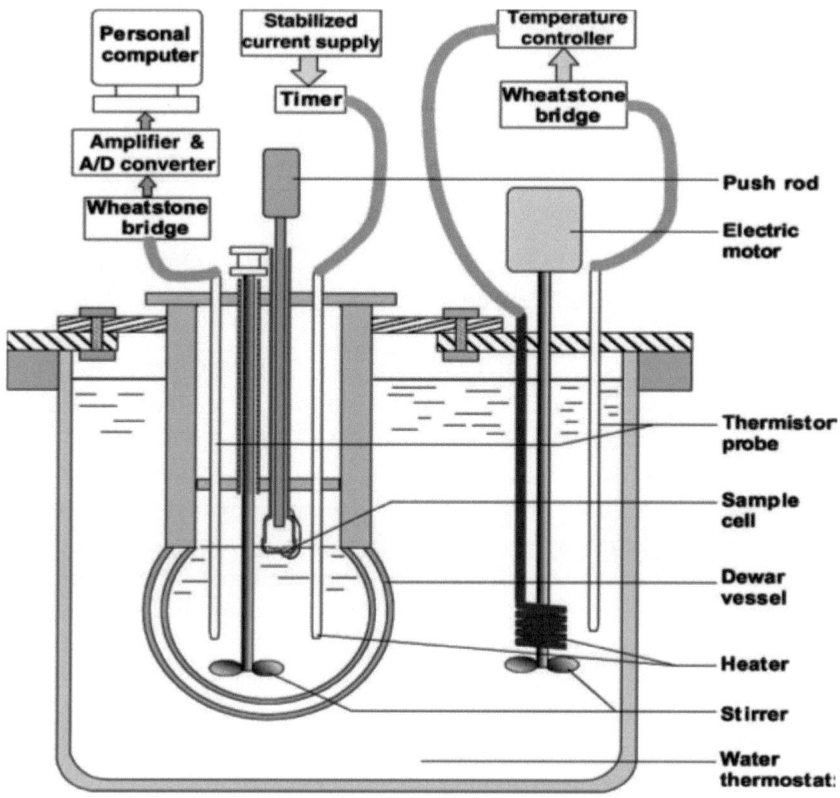

Figure 9. The schematic diagram of isoperibol solution-reaction calorimeter.

the products in above cycle under the same condition was within the range of experimental error and may be omitted because the amount of H_2O (l) was very small according to the stoichiometric number of H_2O (l) in above cycle.

As illustrated in **Figure 10**, the molar enthalpy of formation of $Nd(Gly)_2Cl_3{\cdot}3H_2O(s)$ is given as:

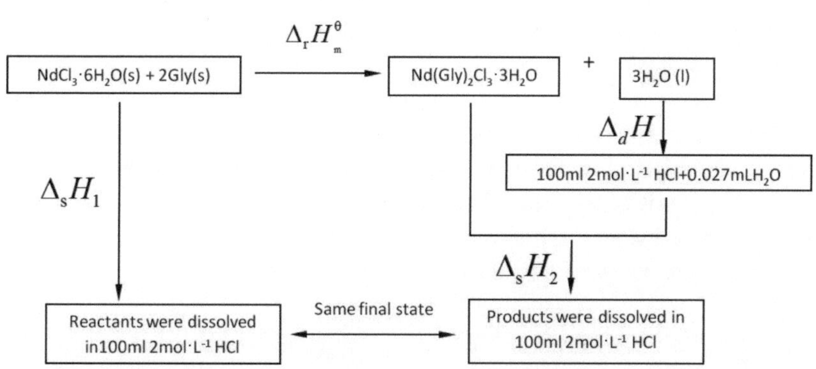

Figure 10. Hess thermochemical cycle.

$$\Delta_r H^\theta{}_m = \Delta_f H^\theta{}_m[\text{Nd(Gly)}_2\text{Cl}_3 \cdot 3\text{H}_2\text{O, s}] + 3\Delta_f H^\theta{}_m(\text{H}_2\text{O, l}) + \Delta_f H^\theta{}_m(\text{NdCl}_3 \cdot 6\text{H}_2\text{O, s}) \\ +2\Delta_f H^\theta{}_m(\text{Gly, s}) = \Delta s H_1 - \Delta s H_2 - \Delta_d H \approx \Delta s H_1 \Delta s H_2 \quad (19)$$

where $\Delta_f H^\theta{}_m$[Nd(Gly)$_2$Cl$_3$·3H$_2$O,s], $\Delta_f H^\theta{}_m$(NdCl$_3$·6H$_2$O,s), and $\Delta_f H^\theta{}_m$(Gly,s) are the molar enthalpy of formation of the corresponding compounds. $\Delta_f H^\theta{}_m$ [Nd(Gly)$_2$Cl$_3$·3H$_2$O,s] could be calculated from the following equation:

$$\Delta_f H^\theta{}_m[\text{Nd(Gly)}_2\text{Cl}_3 \cdot 3\text{H}_2\text{O, s}] = \Delta s H_1 - \Delta s H_2 + \Delta_f H^\theta{}_m(\text{NdCl}_3 \cdot 6\text{H}_2\text{O, s}) \\ +2\Delta_f H^\theta{}_m(\text{Gly, s}) - 3\Delta_f H^\theta{}_m[\text{H}_2\text{O, l}] \quad (20)$$

The mixture of NdCl$_3$·6H$_2$O(s) and Gly(s) at mole ratio of n(NdCl$_3$·6H$_2$O):n(Gly) = 1:2 was dissolved in 100 mL of 2 mol·L^{-1} HCl at 298.15 K. The final solution obtained was named as Solution A. Dissolution enthalpy of 1 mol Nd(Gly)$_2$Cl$_3$·3H$_2$O (s) in 100 mL of 2 mol·L^{-1} hydrochloric acid was determined under the same condition as above. The final solution obtained was named as Solution A′. Finally, UV–vis spectroscopy and the data of the refrangibility were applied to confirm whether Solution A was in the same thermodynamic state as that of Solution A′. These results mentioned above indicated that chemical components and physical-chemistry properties of Solution A were consistent with those of Solution A′. In this chapter, UV–vis spectrum and the data of the refrangibility of Solution A obtained agreed with those of Solution A′. These results demonstrated that the designed thermochemical cycle is reasonable and reliable, hence can be used to calculate the standard molar enthalpy of formation of Nd(Gly)$_2$Cl$_3$·3H$_2$O(s).

3.2. Results and discussion

3.2.1. Thermal analysis

The thermal analysis results of Nd(Gly)$_2$Cl$_3$·3H$_2$O are shown in **Figures 11** and **12**, respectively. The DSC curve (**Figure 11**) reveals that the two endothermic peaks appear at 146.68 and 282.21°C, which are in agreement with those obtained from the TG-DTG (thermogravimetric-differential thermogravimetric) experiments at 147.5 and 283.0°C.

The mass loss of Nd(Gly)$_2$Cl$_3$·3H$_2$O was mainly divided to two steps. Based on the TG curve and the structure of Nd(Gly)$_2$Cl$_3$·3H$_2$O, the conclusion can be drawn that from 59 to 188°C the three coordinating water molecules break away from the complex. The experimental mass loss (%) of the first step was found to be 8.9%. The result is consistent with the theoretical mass loss, 11.9%, of the three water molecules. The second step of the mass loss from 262 to 904°C should be due to the glycine being separated from the Nd^{3+} cations, and the residue of the TG experiment should be NdCl$_3$ and Nd. The experimental mass loss (%) of the decomposition is 52.8%, which is in accord with the theoretical mass loss, 48.6%. The possible mechanism of the thermal decompositions was deduced as

$$3\text{Nd(Gly)}_2\text{Cl}_3 \cdot 3\text{H}_2\text{O} \xrightarrow[8.9\%\ (11.9\%)]{59-188^\circ\text{C}} 3\text{Nd(Gly)}_2\text{Cl}_3 \xrightarrow[52.8\%\ (48.6\%)]{262-904^\circ\text{C}} \text{NdCl}_3 + 2\text{Nd}$$

http://dx.doi.org/10.5772/intechopen.69473

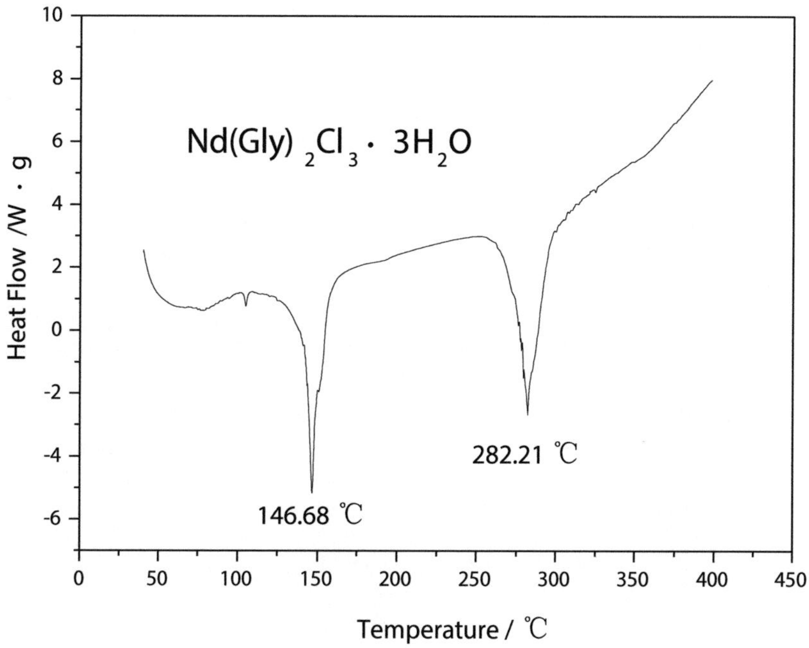

Figure 11. DSC curve of the complex $Nd(Gly)_2Cl_3{\cdot}3H_2O$.

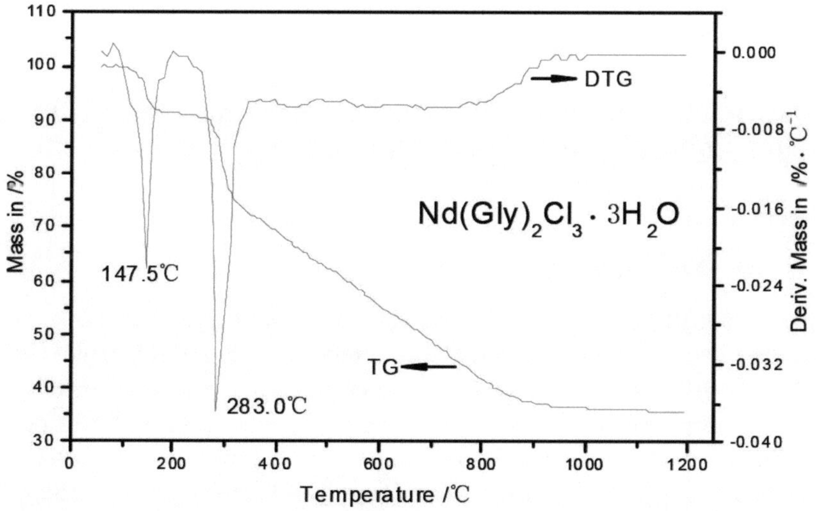

Figure 12. TG-DTG curves of the complex $Nd(Gly)_2Cl_3{\cdot}3H_2O$.

3.2.2. The standard molar enthalpy of formation of the complex $Nd(Gly)_2Cl_3{\cdot}3H_2O$

The experimental values of the dissolution enthalpies of the mixture prepared at the mole ratio of the reactants were measured. Dissolution enthalpies $\Delta_s H_1$ and $\Delta_s H_2$ were determined by the dissolution process tested. The experiment was repeated five times in the same condition, and

No.	1	2	3	4	5	Mean value
$\Delta_s H_1$/kJ·mol^{-1}	4.5298	−4.7254	−4.3564	−4.4486	−4.2986	−4.4718
$\Delta_s H_2$/kJ·mol^{-1}	10.3709	10.3722	10.3699	10.3678	10.3731	10.3707

Table 4. The determination results of the ΔsH_1 and ΔsH_2 at 298.15 K.

the results were listed in **Table 4**. The enthalpy change $\Delta_r H^{\theta}{}_{m}$ could be calculated in accordance with expression (19) as follows:

$$\Delta_r H^{\theta}{}_{m} = \Delta sH_1 - \Delta sH_2 = -4.4718 - 10.3707 = (-14.8425 \pm\ 0.1512)\text{kJ} \cdot \text{mol}^{-1}$$

The experimental values of the dissolution enthalpies of the reactants and products in the thermochemical cycle are combined with the following auxiliary thermodynamic data:

$$\Delta_f H^{\theta}{}_{m}(\text{NdCl}_3 \cdot 6\text{H}_2\text{O, s}) = -2867.5\ \text{kJ} \cdot \text{mol}^{-1}\ [49]$$

$$\Delta_f H^{\theta}{}_{m}(\text{H}_2\text{O, l}) = -285.8\ \text{kJ} \cdot \text{mol}^{-1}\ [49]$$

$$\Delta_f H^{\theta}{}_{m}(\text{Gly, s}) = -528.5\ \text{kJ} \cdot \text{mol}^{-1}\ [50]$$

According to the expression (20), the standard molar enthalpy of formation of the complex Nd (Gly) ^{2}Cl^3·3H^{2}O (s) at 298.15 K can be calculated as follows:

$$\Delta_f H^{\theta}{}_{m}[\text{Nd(Gly)}_2\text{Cl}_3 \cdot 3\text{H}_2\text{O, s}] = (-3081.3 \pm 1.08)\text{kJ} \cdot \text{mol}^{-1}$$

4. Microcalorimetric study on biological effect of rare-earth complex with amino acid: La(Glu)(Im)$_6$(ClO$_4$)$_3$·4HClO$_4$·4H$_2$O

4.1. Experiment

4.1.1. Synthesis and characterization of the complex

The complex, La(Glu)(Im)$_6$(ClO$_4$)$_3$·4HClO$_4$·4H$_2$O, was synthesized in water solution. All the starting materials were analytical reagents from the Beijing Chemical Reagent Co. Rare-earth oxide (La_2O_3) was dissolved in an excess amount of perchloric acid, and the concentration of the solution was determined by EDTA（ethylene diamine tetraacetic acid) titrimetric analysis. Then solid glutamic acid was added to the solution of La^{3+} in molar ratio of La^{3+}:Glu = 1:1. After the pH value of the reaction mixture was carefully adjusted to 6.0 by adding 0.1 mol·L^{-1} NaOH slowly, imidazole was added as the same molar as glutamic acid. The reaction was performed at 60°C in water bath for 5 h. Then the solution was condensed at 80°C for 5 h and put into desiccator in the fridge at −4°C. The crystal was obtained after about one month.

An elemental analysis apparatus (Model PE-2400 II, USA) was used to measure the ratio of C, H, and N in the complex. Content of La % was determined by EDTA titration. The experimental values are: La, (9.423%), C (18.363%), N (12.964%) and H (2.736%), which are close to the theoretical values, La (9.057%), C (18.883%), N (12.452%) and H (2.805%). The sample formula was determined to be $La(Glu)(Im)_6(ClO_4)_3{\cdot}4HClO_4{\cdot}4H_2O$, and the purity obtained from the EDTA titration under the same conditions was found to be 99.79%.

IR spectra were obtained using KBr pellets with a Tensor 27 (Bruker) spectrometer. It can be seen from the IR spectra of the complex that the symmetrical resonance frequencies, $\nu_{s\ (COO-)}$, shifted from 1431 to 1413 cm^{-1}, which suggests that the carboxyl group of Glu has coordinated to the metal ions.

4.1.2. *The* Escherichia coli *(*E. coli*) DH5α*

The *E. coli* DH5*α* was provided by Biomass Conversion Technology Group, Dalian Institute of Chemical Physics, CAS, Dalian 116023, PR China. The strain of *E. coli* DH5*α* was stored in 10% glycerol solution at −20°C.

The *E. coli* DH5*α* used in the present research was prepared as follows. A single colony of *E. coli* DH5*α* from LB(Luria-Bertani) plates was inoculated into a 10 mL LB culture medium and cultivated at 37°C in a rotary shaker (220 rpm) for 12 h under aerobic condition. Then, 200 μL of the above suspension was inoculated into 10 mL LB culture medium at 37°C in a rotary shaker (220 rpm) for 2.5 h once again. The value of optical density (OD) of the *E. coli* DH5*α* suspension was measured to be about 0.6 by spectrophotometry at λ = 600 nm. Per 200 mL of LB culture medium contained 2 g tryptone, 1 g yeast extract powder, 2 g and NaCl (pH 7.0–7.2). It was sterilized in high-pressure steam at 120°C for 20 min before experiment.

4.1.3. *Microcalorimeter*

Thermal Analysis Microcalorimeter (TAM) Air is a type of eight-channel twin isothermal microcalorimeters that originally was developed for the study of the hydration process of cement and concrete. TAM represents an ultra-sensitive heat flow measurement which is complementary to TA instruments differential scanning calorimeters (DSC). "Air" means the thermostat type. However, the high sensitivity and excellent long-term stability make TAM Air useful also for other applications such as stability testing of energetic materials, determining the shelf life of food and monitoring biological processes [51].

A TAM Air Isothermal Microcalorimeter (see **Figure 13**) manufactured by Thermometric AB Company of Sweden (incorporated by TA Instruments in 2006) was used to measure the power-time curves of the metabolism of *E. coli* DH5*α* at 37°C. The main structure of the instrument is eight channels, each of which consists of aluminum heat sink, vessel holder, and thermocouple plate, and each two of them are twin. Glass reaction vessel of 20 mL was used in each channel. The working temperature in the instrument was 5–60°C controlled by air thermostat. The deviation of the controlled temperature is within ±0.02°C. A computer was employed to record the voltage-time signals continuously which were converted to power-time signals through calibration. Thermal power detection limit was stated to be ± 2 μW. The development and theory of many kinds of multi-channel isothermal microcalorimeters have been studied by Wadsö [52, 53].

The ampoule method was used for the microcalorimetric measurement in this work. Luria-Bertani (LB) culture medium (10 mL) containing object compound with different concentrations

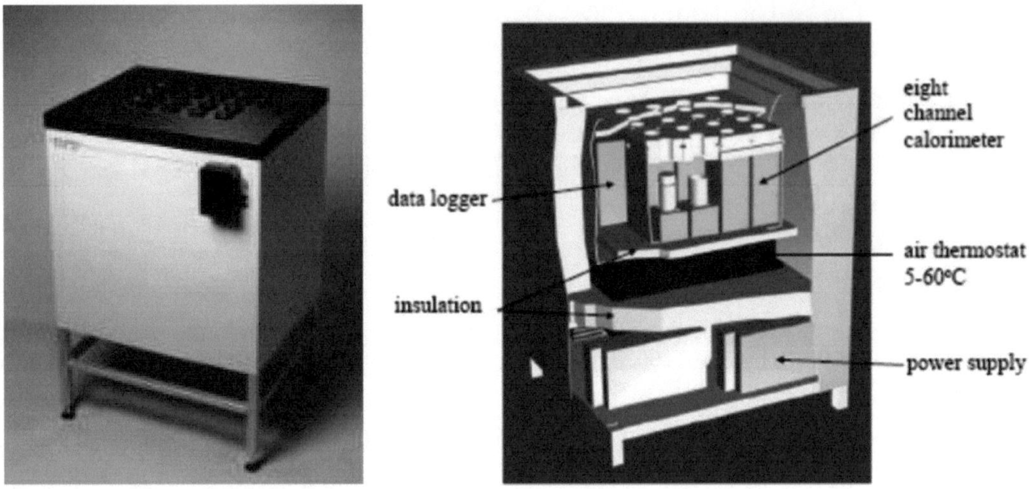

Figure 13. Practicality and cutaway views of TAM Air 8-channel isothermal calorimeter.

was put into eight sample ampoules, which had been cleaned and sterilized. Then, the *E. coli* DH5α suspension was inoculated into the above eight ampoules. At last, the ampoules were sealed with ampulla cap by ampoule filler. The working temperatures of the calorimeter were set and controlled at 37°C. The power-time signals were recorded at intervals of 1 min.

The ampoule method was used for the microcalorimetric measurement in this work. LB culture medium (10 mL) containing object compound with different concentrations was put into eight sample ampoules, which had been cleaned and sterilized. Then, the *E. coli* DH5α suspension was inoculated into the above eight ampoules. At last the ampoules were sealed with ampulla cap by ampoule filler. The working temperatures of the calorimeter were set and controlled at 37°C. The power-time signals were recorded at intervals of 1 min.

4.2. Results and discussion

4.2.1. Thermokinetics

In the log phase of growth, the growth of *E. coli* DH5α cells is exponential [54]. It can be expressed as follows:

$$n_t = n_0 \exp(kt) \tag{21}$$

Where t is the incubation time, n_t is the cell number at time t, n_0 is the initial cell number, and k is the constant of cell growth rate. If the power output of each cell is denoted as Pw, then

$$n_t Pw = n_0 Pw \exp(kt) \tag{22}$$

We define P_0 as the initiative power output and P_t as the power output at time t, then Eq. (2) can be rewritten as follows:

http://dx.doi.org/10.5772/intechopen.69473

$$P_t = P_0 \exp\,(kt) \tag{23}$$

or

$$\ln P_t = \ln P_0 + kt \tag{24}$$

For thermokinetic parameters, inhibitory ratio (I) is defined as:

$$I = [(k_0 – k_c)/k_0]/100\% \tag{25}$$

Where k_0 is the rate constant of the control, and k_c is the rate constant of *E. coli* DH5α growth inhibited by the complex with a concentration of C.

4.2.2. Thermokinetic parameters

Figures 14 and **15** show the growth power-time curves of *E. coli* DH5α under various concentrations (0–5.0 μg·m mL^{-1}) of the complex at 37°C. It can be seen from the figures that growth curve of *E. coli* DH5α can be divided into four typical phases, which are lag phase, log phase, stationary phase, and decline phase. The thermokinetic parameters of growth, growth rate constants (k/10^{-3}·min^{-1}), the maximum heat power (P_m/μW), and the time of the maximum heat power (t_m/min) are derived from the curves by using Eqs. (1)–(5), and the results are listed in **Tables 5** and **6**.

The relationships among the concentration (C), the rate constant (k), and the maximum heat power (P_m) were plotted in **Figure 16** at different concentrations (0, 0.1, 0.5, and 1.0 μg·mL^{-1}) based on the data of **Table 5**. It can be seen from **Figure 16** that the rate constant (k) and the maximum heat power (P_m) were both increasing with the increase of the concentration from

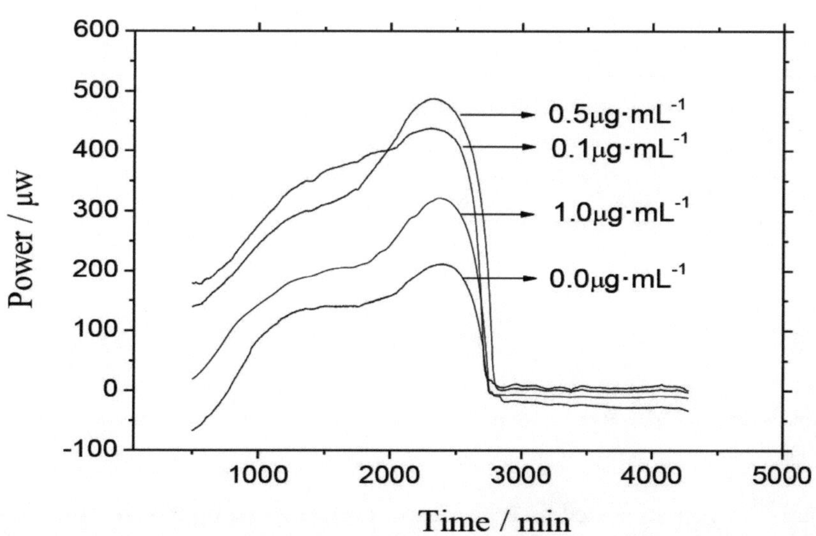

Figure 14. Growth curves of *E. coli* DH5α at different concentrations (0, 0.1, 0.5, and 1.0 μg·mL^{-1}) of the complex at 37°C.

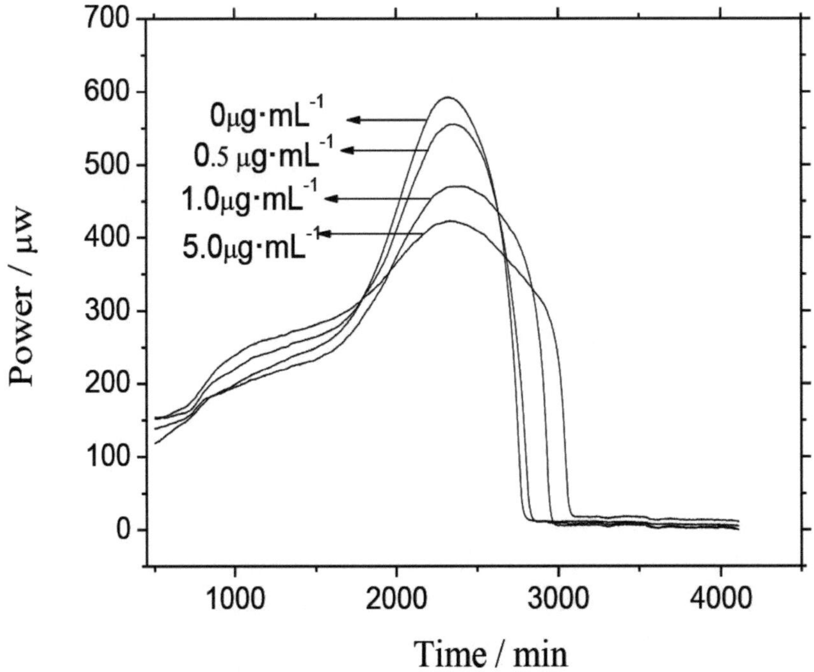

Figure 15. Growth curves of *E. coli* DH5α at different concentrations (0, 0.5, 1.0, and 5.0 μg·mL^{-1}) of the complex at 37°C.

C/μg·mL^{-1}	k/10^{-3}·min^{-1}	R	P_m/μW	t_m/min	I%
0	0.2	0.9992	211.9	2390	0
0.1	0.25	0.9967	438.2	2297	13.3
0.5	0.3	0.9989	488.0	2328	20.0
1.0	0.08	0.9981	379.6	2391	53.3

Table 5. Thermodynamic parameters of growth of *E. coli* DH5α at different concentrations of the complex.

C/μg·mL^{-1}	k/10^{-3} min^{-1}	R	P_m/μW	t_m/min	I%
0	1.5	0.9993	592.3	2308	0
0.5	1.3	0.9961	555.4	2365	25
1.0	1.2	0.9985	470.5	2422	50
5.0	0.7	0.9981	423.8	2332	80

Table 6. Thermodynamic parameters of growth of *E. coli* DH5α at different concentrations of the complex.

0 to 0.5 μg mL^{-1}, which indicated that the complex could speed up the growth of *E. coli* DH5α. When the concentration (C) reached 0.5 μg mL^{-1}, the value of rate constant (k) showed a highest value. It demonstrated that the complex could promote clearly the growth of *E. coli*

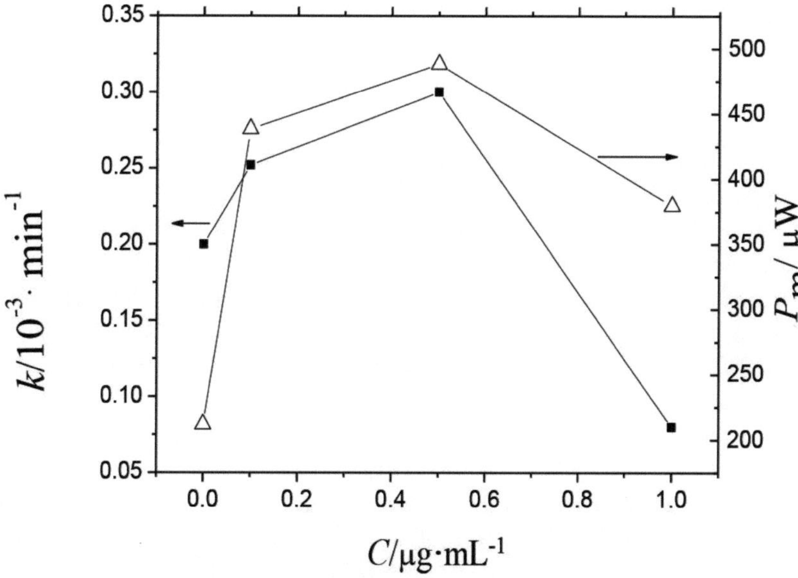

Figure 16. *K-c* curves of *E. coli* DH5α at different concentrations (0, 0.1, 0.5, and 1.0 μg mL^{-1}) of the complex at 37°C.

DH5α at this time. At the concentration (*C*) from 0.5 to 1.0 μg mL^{-1}, the line decreased, which was consistent with **Figure 17**.

Figure 17 was plotted from the data of **Table 6**, when the concentration (*C*) was from 0.5 to 5.0 μg mL^{-1}. It showed that the rate constant (*k*) and the maximum heat power (P_m) were both decreasing with the increase of the concentration (*C*), which indicated that the complex inhibited the growth of *E. coli* DH5α at this range of concentrations (*C*). When the concentration reached 5.0 μg mL^{-1}, *E. coli* DH5α almost could not grow.

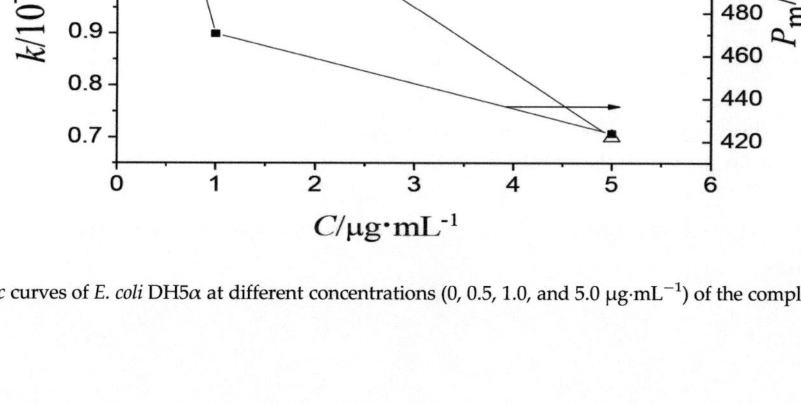

Figure 17. *K-c* curves of *E. coli* DH5α at different concentrations (0, 0.5, 1.0, and 5.0 μg·mL^{-1}) of the complex at 37°C.

5. Conclusions

In this chapter, the following three rare-earth complexes with amino acids were synthesized and characterized by element analysis, infrared（IR） spectrum, and x-ray diffraction (XRD) analysis. The thermodynamic property studies on these complexes were performed. For the first rare-earth complex, $Eu(Glu)(Im)_5(ClO_4)_3{\cdot}3HClO_4{\cdot}6H_2O$, the low-temperature molar heat capacity was measured by adiabatic calorimetry. Two thermal abnormalities interpreted as phase transitions were discovered at 216.18 and 246.86 K by the AC technique, and proved by DSC. The thermodynamic functions $[H_T\text{-}H_{298.15}]$ and $[S_T\text{-}S_{298.15}]$ were derived in the temperature range from 80 to 320 K. The mechanism of thermal decomposition was deduced in terms of the results of TG. For the second rare-earth complex, $Nd(Gly)_2Cl_3{\cdot}3H_2O$, the standard molar enthalpy of formation was determined to be (-3081.3 ± 1.08) $kJ{\cdot}mol^{-1}$ by an isoperibol solution-reaction calorimeter in terms of a designed thermochemical cycle. For the third rare-earth complex, $Eu(Glu)(Im)_5(ClO_4)_3{\cdot}3HClO_4{\cdot}6H_2O$, the microcalorimetry was used to investigate the interaction between the complex and the *Escherichia coli* DH5α to elucidate the biological effects of the complex. The thermokinetic parameters of growth of the *E. coli* DH5α were obtained. The results show that the complex has a stimulatory effect on the growth of *E. coli* DH5α below the concentration 0.5 $\mu g\ mL^{-1}$, and the complex could inhibit its growth in the concentration range between 0.5 and 5.0 $\mu g\ mL^{-1}$.

Acknowledgements

This work was financially supported by the National Natural Science Foundation of China under the grant NSFC No. 20373072 and 21473198. Parts of this chapter are reproduced from authors' recent conference publication, work, etc. [33, 36].

Author details

Zhi Cheng Tan[1]*, Quan Shi[1], Xue Chuan Lv[2] and Bei Ping Liu[3]

*Address all correspondence to: tzc@dicp.ac.cn

1 Thermochemistry Laboratory, Dalian Institute of Chemical Physics, Chinese Academy of Science, Dalian, China

2 School of Chemistry and Material Science, College of Chemistry Chemical Engineering and Environmental Engineering, Liaoning Shihua University, Fushun, China

3 College of Chemistry and Materials Engineering, Hunan University of Arts and Science, Hunan, China

References

[1] Tan ZC, Zhang ZY, Yin JZ, Jiang BG, Sun TS. Low-temperature heat capacities and thermodynamic properties of rare-earth triisothiocyanate hydrates, $RE(NCS)_3$. nH_2O. Chinese Science Bulletin. 1991;**36**(13):1086–1090

[2] Tan ZC, Takasuke M, Hiroshi S, Zhang ZY, Yin JZ, Jiang BG, et al. Low temperature heat capacities and thermodynamic properties of rare earth triisothiocyanate hydrate. 1. La $(NCS)_3 \cdot 7H_2O$ and $Ce(NCS)_3 \cdot 7H_2O$. Science in China Series B-Chemistry. 1992;**35** (4):1391–1403

[3] Tan ZC, Takasuke M, Hiroshi S, Zhang ZY, Yin JZ, Jiang BG, et al. Low temperature heat capacities and thermodynamic properties of rare earth triisothiocyanate hydrates. 2. Pr $(NCS)_3$ $7H_2O$ and $Nd(NCS)_3$ $7H_2O$. Science in China Series B-Chemistry. 1996;**39**(3):276–284

[4] Tan ZC, Takasuke M, Hiroshi S, Zhang ZY, Yin JZ, Jiang BG, et al. Low-temperature heat capacities and thermodynamic properties of rare-earth triisothiocyanate hydrates 3. Sm $(NCS)_3 \cdot 6H_2O$, $Gd(NCS)_3 \cdot 6H_2O$, $Yb(NCS)_3 \cdot 6H_2O$ and $Y(NCS)_3 \cdot 6H_2O$. Science in China, Series B-Chemistry. 1997;**40**(2):165–176

[5] Anghileri LJ. On the antitumor activity of gallium and lanthanides. Arzneim-Forsch. 1975;**25**:793–795

[6] McCarthy GJ. Rare Earths in Modern Science and Technology. Vol. 2. Plenum Press; New York, 1980. pp. 25–105

[7] Sudhindra NM, Joshi GK, Bhutra MP. Syntheses and absorption spectral studies of Praseodymium (II) and Neodymium (II) complexes with amino acids. Indian Journal of Chemistry. 1982;**21A**:275–278

[8] Glowiak T, Legendziewicz J, Huskowska E, Gawryszewska P. Ligand chirality effect on the structure and its spectroscopic consequences in $[Ln_2(Ala)_4(H_2O)_8](ClO_4)_6$. Polyhedron. 1996;**15**:2939–2947

[9] Zhang H, Feng J, Zhu WF. Rare-earth element distribution charcteriatics of biological chains in rare-earth element-high background regions and their implications. Biological Trace Element Research. 2000;**73**:19–27

[10] Xu H, Chen L. Study on the complex site of L-tyrosine with rare-earth element Eu^{3+}. Spectrochimica Acta Part A. 2003;**59**:657–662

[11] Li GT, Zhang DS, Li L, Tan ZC, Wu XM, Meng SH. Low-temperature heat capacity and thermochemical study of $Re_2(Pro)_6(H_2O)_4$ $(ClO4)_6$ (Re = Nd, Gd; Pro = proline) crystals. Thermochimica Acta. 2001;**375**(1–2):125–130

[12] Wu XM, Liu Y, Li L, Liu P, Tan ZC, Qu SS. Low-temperature heat capacity and standard enthalpy of formation of samarium glycine perchlorate complex $Sm_2(Gly)_6(H_2O)_4$ $(ClO4)_6$. $5H_2O$. Acta Chimica Sinica. 2001;**59**(9):1396–1400

[13] Liu BP, Tan ZC, Nan ZD, Liu P, Sun LX, Xu F. Calorimetric study and thermal analysis of $[Ho_2(Ala)_4(H_2O)_8]Cl_6$ and $[ErY(Ala)_4(H_2O)_8]$ $(ClO4)_6$. Acta Physico-Chimica Sinica. 2002;**18**(6):481–485

[14] Liu BP, Zhao XS, Li L, Sun LX, Tan ZC. Low-temperature heat capacity and thermochemical study of crystalline $[Y_2(Ala)_4(H_2O)_8]$ $(ClO_4)_6$ (Ala=$CH_3CH(NH_3^+)(COO^-)$). Thermochimica Acta. 2002;**389**(1-2):59–64

[15] Liu BP, Tan ZC, Zhang DS, Nan ZD, Sun LX, Xu F, et al. Low-temperature heat capacity and thermodynamic properties of crystalline $[Re_2(Ala)_4$ $(H_2O)_8]$ $(ClO4)_6$ (Re = Eu, Er; Ala = alanine). Thermochimica Acta. 2002;**390**(1–2):31–37

[16] Lan XZ, Tan ZC, Liu BP, Nan ZD, Sun LX, Xu F. Low-temperature heat capacity and thermal decomposition of crystalline $[Ho(Thr)(H_2O)_5]Cl_3$. Chinese Journal of Chemistry. 2003;**21**(8):1043–1046

[17] Lan XZ, Tan ZC, Liu BP, Nan ZD, Sun LX, Xu F. Low-temperature heat capacity and thermodynamic properties of $[Re_2(Ile)_4(H_2O)_8$ $(ClO_4)_6$ (Re = Nd, Er, Ile = isoleucine). Thermochimica Acta. 2003;**402**(1-2):183–191

[18] Liu BP, Tan ZC, Lan XZ, Yu HG, Zhang DS, Sun LX. Calorimetric study and thermal analysis of $[Gd_{4/3}Y_{2/3}(Gly)_6(H_2O)_4]\cdot(ClO_4)_6\cdot 5H_2O$ and $[ErY(Gly)_6(H_2O)_4]$ $(ClO_4)_6\cdot 5H_2O$ (Gly = glycin). Thermochimica Acta. 2003;**401**(2):233–238

[19] Liu BP, Tan ZC, Lu JL, Lan XZ, Sun LX, Xu F, et al. Low-temperature heat capacity and thermodynamic properties of crystalline $RE(Gly)_3(H_2O)_2$ $Cl_3\cdot 2(H_2O)$ (RE = Pr, Nd, Gly = Glycine). Thermochimica Acta. 2003;**397**(1-2):67–73

[20] Liu BP, Tan ZC, Nan ZD, Liu P, Sun LX, Xu F, et al. Calorimetric study and thermal analysis of $[ErY(Ala)_4(H_2O)_8](ClO_4)_6$ (Ala=alanine). Journal of Thermal Analysis and Calorimetry. 2003;**71**(2):623–628

[21] Liu BP, Tan ZC, Yu HG, Lan XZ, Zhang DS, Liu P, et al. Thermodynamic properties of $[HO_2(Ala)_4(H_2O)_8]Cl_6$ (Ala = alanine). Acta Physico-Chimica Sinica. 2003;**19**(5):445–449

[22] Yu HG, Liu Y, Tan ZC, Dong JX, Zou TJ, Huang XM, et al. A solution-reaction isoperibol calorimeter and standard molar enthalpies of formation of $Ln(hq)_2Ac$ (Ln = La, Pr). Thermochimica Acta. 2003;**401**(2):217–224

[23] Lan XZ, Tan ZC, Liu BP, Nan ZD, Sun LX, Xu F. Low-temperature heat capacity and thermal decomposition of crystalline $[Er_2(His\cdot H+)(H_2O)_8]$ $(ClO_4)_6\cdot 4H_2O$. Thermochimica Acta. 2004;**416**(1-2):55–58

[24] Wu XM, Song YJ, Tan ZC, Qu SS. Measurement of low-temperature heat capacity and standard enthalpy of formation of praseodymium glycin perchlorate complex. Chinese Journal of Chemical Physics. 2005;**18**(2):284–288

[25] Di Y-Y, Tan ZC, Li Y-S. Low-temperature heat capacities and thermochemistry of the complex of praseodymium perchlorate with L-alpha-glutamic acid: $[Pr_2(Glu)_2(ClO_4)$ $(H_2O)_7](ClO_4)_3\cdot 4H_2O$. Acta Chimica Sinica. 2006;**64**(13):1393–1401

[26] Wu XM, Tan ZC, Lv XC, Qu SS. Synthesis, heat capacity and enthalpy of formation of $[Ho_2(L\text{-}Glu)_2(H_2O)_8]$ $(ClO_4)_4{\cdot}H_2O$. Thermochimica Acta. 2006;**441**(2):199–202

[27] Wu XM, Qu CS, Qu SS, Tan ZC. Low-temperature heat capacity and standard entralpy of formation of neodymium glycine perchlorate complex $[Nd_2(Gly)_6(H_2O)_4(ClO_4)_6{\cdot}5H_2O$. Journal of Thermal Analysis and Calorimetry. 2007;**90**(2):569–573

[28] Zhang JN, Tan ZC, Liu BP, Shi Q, Tong B. Low-temperature heat capacities and thermochemistry properties of $Sm(Val)Cl_3{\cdot}6H_2O$. Acta Physico-Chimica Sinica. 2008;**24**(8):1378–1382

[29] Lv X-C, Tan ZC, Gao X-H. Synthesis and thermochemical properties of new ternary lanthanum complex $La(Glu)(Im)_6(ClO_4)_3{\cdot}4HClO_4{\cdot}4H_2O$. Acta Physico-Chimica Sinica. 2009;**25**(10):1945–1950

[30] Wu XM, Li W, Tan Z-C, Qu SS. Heat capacity and thermodynamical properties of the crystal of $[RE_2(Glu)_2(H_2O)_8]$ $(ClO_4)_4{\cdot}H_2O$ (RE = Nd, Eu, Dy). Science in China Series B-Chemistry. 2009;**52**(7):862–867

[31] Zhang JN, Wang H, Tan ZC, Liu BP, Shi Q, Tong B. Low-temperature heat capacities of crystalline $Ho(Gly)_3Cl_3{\cdot}3H_2O$ from 78 to 348 K. Journal of Rare Earths. 2009;**27**(6):919–922

[32] Zhang JN, Tan ZC, Liu BP, Shi Q, Tong B. Thermochemical Behavior of Crystalline RE $(Val)Cl_3{\cdot}6H_2O$ (RE = Nd, Er, Val = Valine). Journal of Chemical and Engineering Data. 2009;**54**(2):392–395

[33] Lv XC, Tan ZC, Gao XH, Yang LN. Microcalorimetric study about biological effect of a synthetic complex: $La(Glu)(Im)_6(ClO_{4)3}{\cdot}4HClO_4{\cdot}4H_2O$. Thermochimica Acta. 2010;**510**:213–216

[34] Gao XH, Xu P, Duan WC, Lv XC, Tan ZC, Lu Q. Low-Temperature heat capacity and thermodynamic functions of $Ho(NO_3)_3(C_2H_5O_2N)_4{\cdot}H_2O$. Acta Physico-Chimica Sinica. 2013;**29**(10):2123–2128

[35] Liu B-P, Tan ZC. Thermodynamic properties of $Nd(Gly)_2Cl_3{\cdot}3H_2O$ and $Pr(Ala)_3Cl_3$ H_2O. Acta Physico-Chimica Sinica. 2013;**29**(1):17–22

[36] Zhang DS, Tan ZC, Liu BP, Huang XB. Thermodynamic study on complex of neodymium with glycine. Journal of Chemical Thermodynamics. 2015;**83**:27–32

[37] Luo GH, Gao XH, Pan L, Lv XC, Tan ZC., Low-temperature molar heat capacities and thermodynamic properties of a new rare earth complex Er_2(mu 2-Gly)$_6(H_2O)_4{\cdot}Na_2(ClO_4)_8$ $(H_2O)_2{\cdot}4H_2O$. Journal of Thermal Analysis and Calorimetry. 2016;**126**(2):871–879

[38] Pan L, Gao XH, Lv XC, Tan ZC. Low-temperature molar heat capacity and thermodynamic properties of rare earth complex. $[EuCu_6$(mu-OH) $_3$(Gly) $_6Im_6](ClO_4)_6{\cdot}3H_2O$. Journal of Thermal Analysis and Calorimetry. 2016;**124**(1):429–435

[39] Nakamoto K. Infrared Spectra of Inorganic and Coordination Compounds. 4th ed. New York: John Wiley & Sons Inc; 1986. p. 258

[40] Wayda AL, Kaplan ML. Mixed ligand imidazole complexes of organolanthanides. Polyhedron. 1990;**9**:751–756

[41] Tan ZC, Shi Q, Liu BP, Zhang HT. A fully automated adiabatic calorimeter for heat capacity measurement between 80 and 400 K. Journal of Thermal Analysis and Calorimetry. 2008;**92**(2):367–374

[42] Donald GA. Thermodynamic properties of synthetic sapphire standard reference material 720 and the effect of temperature- scale difference on thermodynamic properties. Journal of Physical and Chemical Reference Data. 1993;**22**:1441–1452

[43] Yukawa Y, Igarashi S, Masuda Y, Oguni M. Phase transition and glass transition concerning configurational order/disorder of ions in crystalline $(TMA)_2[Sr\{Ni(Pro)_2\}_6](ClO_4)_4$ and $(TMA)[Sm\{Ni(Pro)_2\}_6](ClO_4)_4$. Journal of Molecular Structure. 2002;**605**:277–290

[44] Anna MM, Edward M, Łukasz H, Ireneusz N, Ewa Ś, Jacek Ś, Stanisław W. Phase transition, molecular motions, structural changes and low- frequency vibrations in $[Cu(NH_3)_5](ClO_4)_2$. Chemical Physics. 2005;**317**:188–197

[45] Hangam SS, Westrum ER. Heat capacities and thermodynamic properties of globular molecules I. adamantane and hexamethylenetetramine. Thermodynamic Properties of globular Molecules. 1960;**64**:1547–1551

[46] Udowenko AA, Laptash NM, Maslennikova IG. Orientation disorder in ammonium elpasolites. Crystal structures of $(NH_4)_3AlF_6$,$(NH_4)_3TiOF_5$ and $(NH_4)_3FeF_6$. Journal of Fluorine Chemistry. 2003;**124**:5–15

[47] Di YY, Tan ZC, Zhang GQ, Chen SP, Liu Y, et al. Low-Temperature heat capacity and standard molar enthalpy of formation of the complex $Zn(Thr)SO_4H_2O(s)$. Thermochimica Acta. 2003;**400**(1-2):43–49

[48] Liu YP, Di YY, He DH, Kong YX, Yang WW, Dan WY. The Journal of Chemical Thermodynamics. 2010;**42**:513–517

[49] Sober HA. CRC Handbook of Biochemistry. Ohio: Chemical Rubber Co; 1971. B-63

[50] Cox JD, Wagman DD, Medvedev VA. CODATA Key Values for Thermodynamics. New York: Hemisphere Publishing Corp; 1989

[51] 2009 TA Instruments. http://www.tainstruments.com/

[52] Barros N, Feijóo S, Simoni A, Critter SAM, Airoldi C. Interpretation of the metabolic enthalpy changes calculated for microbiol growth reactions in soils J. Therm. Anal. Calorim. 2001;**63**:577–588

[53] Wadsö I. Thermochim. Acta 2002;**394**:305–311

[54] Xie CL, Tang HK, Qu SS. Microcalorimetric study of bacterial growth. Thermochimi. Acta 1988;**123**:33–41

Section 4

Ferroelectric Films Doped with Rare Earths

Chapter 6

The Effects of Lanthanum Dopant on the Structural and Optical Properties of Ferroelectric Thin Films

Irzaman, Hendradi Hardhienata,
Akhiruddin Maddu, Aminullah and Husin Alatas

Additional information is available at the end of the chapter

http://dx.doi.org/10.5772/intechopen.69029

Abstract

This chapter presents the effect of lanthanum (La) dopant on the structural and optical properties of the following ferroelectric thin films: barium strontium titanate ($Ba_{0.5}Sr_{0.5}TiO_3$) (BST), lithium tantalate ($LiTaO_3$), and lithium niobate ($LiNbO_3$). We applied X-ray diffraction, Fourier transform infrared spectroscopy (FTIR), and in some cases, atomic force microscopy to investigate the structural properties, functional groups, and particle size, as well surface roughness of these lanthanum-doped thin films, respectively, whereas ultra violet-visible (UV-Bis) spectrometer was applied to investigate the optical bandgap of the La-doped thin films. The results are in agreement with other previously researcher studies using different substrate materials. In general, the effect of La dopant highly affects both the structural and optical properties, changing significantly the grain size, surface roughness, and energy gap, and in certain cases, such as BST doped by La can change the material electric properties from insulator to semiconductor. Therefore, La-doped thin films may offer promising applications in the future.

Keywords: lanthanum dopant, optical properties, structural properties, thin films

1. Introduction

It is well known that lanthanum—denoted by the symbol La and atomic number 57—is one of the most abundant rare earth elements with the physical characteristics of being silvery white metallic. Lanthanum can be found in some rare-earth minerals, usually in combination with cerium and other rare earth elements. According to the periodic table, lanthanum belongs to the group 3 element and is a lanthanide. It is one of the most important members in the rare earth family that can be found in combination with other rare earth elements such as cerium.

The element lanthanum consists of 57 electrons with the electronic configuration $1s^22s^22p^6$ $3s^23p^63d^{10}4s^24p^64d^{10}$ $5s^25p^65d^16s^2$ which can also be written more compactly as [Xe] $5d^16s^2$. Therefore, the atoms has three valence electrons outside the noble gas core, one occupying the 5d and two the 6s subshells forming a +3 oxidation state. The electronic structure of lanthanum determines its physical properties. In this regard, the 4f states play an important role due to its effect on the low melting point and superconducting transition temperature and as has been studied by Smith et al. [1] using an electron excited X-ray appearance potential spectrum (EXAPS). It has since been known that the 4f states in the rare earth elements are found to be significantly higher (e.g., up to 5 eV) in the solid state than in the free atomic state. Nevertheless, the extensive overlap between the 4d and 4f subshells, which is dominated by multiplet splitting, makes it unsuitable for the determination of any single-particle energy level; therefore, other methods such as X-ray-excited electron appearance potential spectroscopy (XEAPS) are used to study its band structure, especially the 4f states of La have been performed by Kanski et al. [2]. Some important physical properties of La are presented in **Table 1**.

There are many applications of lanthanum; its ability as a catalyst is among these. In fact, La can be used as additives to improve the performance of other catalysts. In addition, La can also be applied as additives in glass, carbon lighting for studio lighting and projection, ignition element in lighters and torches, electron cathode, and scintillators [3]. Recently, interest in lanthanum as a dopant or an additive in ferroelectric thin film has increased due to its ability to significantly modify the optical and structural properties of its host material. Ferroelectric materials such as barium titanate ($BaTiO_{3)}$), lead titanate ($PbTiO_3$, lead zirconate ($PbZrO_3$), barium strontium titanate (BST), lithium tantalate ($LiTaO_3$), and lithium niobate ($LiNbO_3$) have many interesting applications in modern day thin film technology, such as photovoltaic thin films, color sensors, and thin film-based light emitting diodes (TFLED). Barium titanate, for example, is one of the most common ferroelectric materials, and it is used as a capacitor and ferroelectric memory due to its excellent dielectric and ferroelectric properties [4].

Physical properties	Value
Crystal structure	Double hexagonal close packed
Electronic configuration	[Xe]$5d^16s^2$
Atomic radius	187 pm
Ionization energy (1st)	538 kJ/mol
Thermal conductivity	13.4 W/(m K)
Magnetic property	Paramagnetic
Density near room temperature (solid)	6.16 g/cm^3
Density at melting point (liquid)	5.94 g/cm^3
Melting point	920°C
Boiling point	3464°C
Molar heat capacity	27.1 J/(mol K)

Table 1. Some physical properties of lanthanum.

http://dx.doi.org/10.5772/intechopen.69029

Ferroelectric materials are known for their ability of spontaneous electric polarization, which can be readily reversed using an external electrical field. Ferroelectricity is a result of the electric dipole moments parallel ordering without the presence of an external/applied electrical field. The adjacent dipoles have the tendency to align themselves parallel to each other forming domain structures with strong magnetic orientation. Many ferroelectric materials such as barium titanate ($BaTiO_3$), lead zirconate titanate (PZT), and lead lanthanum zirconate titanate (PLZT) are having the perovskite-type structure ABO_3.

Moreover, BST has an interesting physical property due to substitution of the barium or titanium ion with small ion/atomic dopant such as La—which has an atomic radius close to Sr^{2+} —may result in structure and microstructure changes. This leads to significant changes in the dielectric and ferroelectric properties of the thin film. The capability of these perovskite structures to host different sized ions in the BT lattice is remarkable and may even cause ferroelectric transition as we will show later in this work. At room temperature, pure barium titanate can be categorized as an electrical insulator. The presence of dopants in the BT lattice could cause lowering of electrical resistivity of materials due to atomic substitution and depend significantly on the concentration of the dopant itself. In some cases, addition of dopants may change the material property from insulating to semiconductor but may well reverse to insulator if a critical threshold is reached.

2. Effect of lanthanum doping on ferroelectric host material

The effects of lanthanum content on other ferroelectric materials such as transparent ferroelectric ceramics have been studied thoroughly by Falcao et al. [5]. In their work, the thermo-optical properties of $(Pb, La)(Zr, Ti)O_3$ or PLZT due to lanthanum addition were studied using thermal lens spectrometry, optical interferometry, and the thermal relaxation calorimetry. PLZT plays an important role in the photonic area, because they can be used in diode pumped laser as an active medium. Using three different methods, they were able to determine the thermal diffusivity, thermal conductivity, changes in the optical path length, and temperature coefficient of the refractive index. They found that lanthanum addition in PLZT causes significant alteration of the mentioned thermo-optical property. However, no significant variation in the specific heat of PLZT as a result of La doping was observed. The highest values of thermal diffusivity, thermal conductivity at room temperature was obtained for a 10 mol% of lanthanum concentration addition, thus they claim that PLZT doped with La have many interesting laser applications, due to its very good thermo-optical properties.

More recently, the effects of La doping on $0.96[Bi1/2(Na_{0.84}K_{0.16})_{1/2}]_{(1-x)}La_xTiO_3$-$_{0.04}SrTiO_3$ (BNKTLa$_x$-ST) crystal structure and piezoelectric properties were investigated by Tran et al. [6], where they show using X-ray diffraction (XRD) that a phase transition from coexistence rhombohedral-tetragonal to a pseudocubic phase occurs due to lanthanum doping, namely, for $x = 0.02$, in the above-mentioned crystal structure. In addition, they found that by investigating the dielectric properties upon La addition, a ferroelectric to ergodic relaxor phase transition was observed.

The other work by Han et al. shows that La-modified bismuth titanate (BLT) can also be doped further by iron atoms, leading to large improvements in the bandgap property even though challenges still exist due to the difficulty in creating BLT without forming a $BiFeO_3$ secondary phase [7]. To remove the secondary phase, Han et al. performed a solid reaction at various calcination temperatures (300–900°C). Furthermore, they apply X-ray diffraction and ultraviolet-visible absorption spectroscopy to investigate the structural and optical properties, respectively. The substitution of Fe atoms can decrease the optical bandgap of BLT, which opens the possibility for a tunable bandgap material.

Although in our work, we applied the chemical solution deposition (CSD) method, it is interesting to compare our result with the work of Bae et al. who present the ferroelectric properties of lanthanum-doped bismuth titanate thin films grown by a sol-gel method [8]. In their work, lanthanum-doped bismuth titanate, $Bi_{3.25}La_{0.75}Ti_3O_{12}$ (BLT), and thin films were grown on Pt(1 1 1)/Ti/SiO_2/Si (1 0 0) substrates by a sol-gel spin coating process. The substrate is then sufficiently annealing at 550–700°C to obtain crystallized thin films. Using X-ray diffraction (XRD), they were able to show that the ratio of the c-axis-oriented La-doped BLT grains significantly depends at the annealing temperature. They also claimed, based on the results, that lanthanum-doped BLT has the potential as an environmentally safe lead-free ferroelectric material, possessing remarkable ferroelectric properties.

Pure barium titanate possesses the property as an electrical insulator at room temperature. The addition of dopants can alter the electrical property of such materials, e.g., altering the energy bandgap, which opens promising application both in the field of optics and electronics. For the barium titanate case, the addition of dopants causes a decrease in the material electrical resistivity. Thus, by adding dopant concentration, one can change the electrical property of certain material from insulating to semiconductor. However, after reaching a certain critical threshold dopant levels, the semiconductor material may become an insulator again.

In the proceeding chapters, we describe the optical and structural properties of lanthanum-doped thin films that were characterized by using UV-visible spectrometer and X-ray diffraction, respectively. The functional groups and particle size were observed by using Fourier transform infrared (FTIR) spectrophotometer and particle size analyzer (PSA). The surface roughness was investigated using 3D (three-dimensional) atomic force microscopy (AFM). We start by explaining the experimental methods and preparation of the substrate as well as thin film growth.

3. Experimental methods, substrate preparation, thin film growth, and characterization techniques

Several methods exist in synthesizing the above-mentioned ferroelectric materials, such as pulsed laser deposition (PLD), metal organic solution deposition (MOSD), reactive sputtering, ion beam-assisted deposition (IBAD), and chemical solution deposition (CSD) method. In most of our work, ferroelectric thin films, e.g., barium titanate ($BaTiO_3$), lead barium strontium titanate (BST), lithium tantalate ($LiTaO_3$), and lithium niobate ($LiNbO_3$), were doped with lanthanum and using the CSD method. This method is based on deposition of chemical solution on the top of a substrate often followed by rotation at certain rotational speed using spin coating. Several advantages of using CSD are as follows: processing temperature can be relatively low, composition homogeneity is achieved, precise control of the composition, as well as large area deposition [9–11].

3.1. Preparation and characterization of BST, BGST, and BTST thin films

Barium strontium titanate ($Ba_{0.5}Sr_{0.5}TiO_3$-BST) was produced by using the chemical solution deposition (CSD) method with ultrasonic mixing of 0.160 g barium acetic + 0.131 g strontium acetic + 0.355 g titanium isopropoxide + 0.060 g gallium oxide as a precursor [9] for 2 h. By mixing barium acetate [Ba ($(CH_3COO)_2$, 99%], strontium acetate [$Sr(CH_3COO)_2$, 99%], and titanium isopropoxide [$Ti(C_{12}O_4H_{28})$, 97.99%] with 2.5 ml of 2-methoxy ethanol as a solvent, a BST solution was obtained. The barium strontium titanate (BGST) thin films were fabricated also via CSD methods with ultrasonic mixing of 0.160 g barium acetic + 0.131 g strontium acetic + 0.355 g titanium isopropoxide + 0.060 g gallium oxide as a precursor in 1.25 ml 2-methoxyethanol solution for 2 h. Meanwhile, barium strontium titanate (BTST) was produced with the same method using 0.160 g barium acetic + 0.131 g strontium acetic + 0.355 g titanium isopropoxide + 0.060 g tantalum oxide as a precursor in 1.25 ml 2-methoxyethanol as solvents. The solutions were then stirred with a Branson 2210 ultrasonic stirrer for 90 min. Afterward, 10% gallium oxide–doped barium strontium titanate (BGST) and 10% tantalum oxide–doped barium strontium titanate (BTST) were prepared on a p-type Si (1 0 0) substrate using the chemical solution deposition (CSD) method with a precursor of 1.00 M. After 20 min in room temperature, the clear solution acquired a milky appearance. To produce a homogenous thin film, we use a spin coating method with a spinning speed of 3000 rpm for 30 s. After the deposition process, the thin film substrate was annealed in a furnace Model Nabertherm Type 27 at 200, 240, and 280°C (low temperature) for 1 h using oxygen gas as its atmosphere.

The surface roughness and grain size analysis of the grown thin films were described by the atomic force microscope (AFM) method at a 5000 × 5000 nm sample area. The rms surface roughness of our BGST thin films with 5000 × 5000 nm area are 0.632, 0.564, and 0.487 nm at the annealing temperature of 200, 240, and 280°C, respectively, whereas the grain size (mean diameter) were 238.4, 219.0, and 185.1 nm at annealing temperature of 200, 240, and 280°C, respectively. The increase of annealing temperature from 200 to 280°C resulted in a decrease of the rms roughness and grain size.

3.2. Preparation and characterization of BSLT thin films

BST doped by lanthanum oxide La_2O_3 or in short, BSLT was obtained by mixing solutions of [$Ba(CH_3COO)_2$, 9%], strontium acetate [$Sr(CH_3COOH)_2$, 99%], titanium isopropoxide [$Ti(C_{12}O_4H_{28})$, 99%], lanthanum oxide [(La_2O_3), 99%], and ethylene glycol as a solvent. The molar fraction of Ba was chosen as 0.55, while the molar fraction of Sr at 0.45 mol/l with 1.25 ml of ethylene glycol solvent. The composition and mass of each solvent were determined using the stoichiometric calculations:

$$0.55\,\mathrm{Ba}\left(\mathrm{CH_3COO}\right)_2+0.45\mathrm{Sr}\left(\mathrm{CH_3COO}\right)_2+\mathrm{Ti}\left(\mathrm{C_{12}O_4H_{28}}\right)+22\mathrm{O_2}\rightarrow \mathrm{Ba_{0.55}Sr_{0.45}TiO_3}+17\mathrm{H_2O}+16\mathrm{CO_2} \quad (1)$$

Meanwhile, barium acetate, strontium acetate, and titanium isopropoxide were obtained by mixing ethylene glycol solvent and vibrated for 1 h using a Branson ultrasonic 2210. Furthermore, using the same method, 6% dopant lanthanum oxide was added and vibrated for 30 min. From the mixing process, we obtained BSLT.

To produce BSLT with 6% doping, a Si (1 1 1) type-p which had been cut with area 1 cm^2 was grown using a spin coating technique at 3000 rpm for 30 s. The next process was optimization of the film by annealing at 850°C for 15 h with a heating rate of 2°C/min using furnace VulcanTM 3-13 and dropped in temperature with the same rate to return to room temperature for 20 min. Afterward, the mass of the sample was weighed. The thin films' thicknesses were determined by the volumetric method.

The optical properties of BSLT were characterized using a Vis-NIR spectrophotometer to see the absorption spectra of thin films in large area waves (380–900 nm), then the value of energy gap can be determined with a Tauc plot and the refractive index, *n* with

$$n = \frac{1+\sqrt{R}}{1-\sqrt{R}} \quad (2)$$

where *R* is the reflectance.

The model of crystal structure of the films that has been made was characterized using X-ray diffraction (XRD), and ICDD database was used to match the diffraction peaks.

3.3. Preparation and characterization of $LiNbO_3$ thin films

$LiNbO_3$ thin films were produced in two stages [10]: in the first stage, the substrate was prepared by cutting the p-type Si (1 0 0) waver with a surface area of 8 × 8 mm then cleaned by aqua bidest and dried. In the second stage, the $LiNbO_3$ powder (precursor) was produced. The method and thin film preparation of lithium niobate were analogous with the previous BST, BGST, and BTST sample, namely, using CSD, so we will not repeat the description in length here. We used three prepared precursors, namely, undoped precursor, 5% lanthanum-doped precursor, and 10% lanthanum-doped precursor. The precursors were obtained by mixing $LiCH_3COO$ powder, Nb_2O_5, with lanthanum, and then dissolve the mixture in 2.5 ml 2-methoxy methanol. The mixing process was conducted using an ultrasonic of Branson 2210 for 90 min and then deposited on the substrate by using a spin coating method at a speed of 3000 rpm two times. The next step was annealing process using the furnace of Vulcan^{TM}3–130. The annealing process for each substrate was started from room temperature with the increasing rate of 1.7°C/min to temperature of 550°C and then held constantly for 8 h before cooled until room temperature.

The optical properties of thin films were characterized by using USB 2000 spectrometer, and their crystal structures were observed using XRD-GBC EMMA. The functional groups and particle size were characterized by using FTIR and PSA, respectively.

3.4. Preparation and characterization of $LiTaO_3$ thin films

The preparation of $LiTaO_3$ thin films was performed first by cutting the (0 0 1) Si substrate with a size of 8 × 8 mm. Then, the substrates were cleaned using aqua bidest and dried. In this case, three $LiTaO_3$ solutions were prepared as in our previous work using the CSD (chemical solution deposition) method [11]. The first solution was prepared by mixing 0.1650 g of $LiCH_3COO$ and 0.5524 g of Ta_2O_5, which were soluble in 2.5 ml of 2-methoxy methanol. For the La-doped substrate, the second solution was prepared by mixing 0.1650 g of $LiCH_3COO$ and 0.5524 g of Ta_2O_5 which were soluble in 2-methoxy methanol to become $LiTaO_3$ 2.5 ml with the addition of 0.0295 g of La_2O_3 as a dopant. The obtained product is a 5% lanthanum-doped $LiTaO_3$ solution. Afterward, the third solution was prepared by mixing 0.1650 g of $LiCH_3COO$ and 0.5524 g of Ta_2O_5 which were soluble in 2.5 ml of 2-methoxy methanol with the addition of 0.0590 g of La_2O_3. The obtained product is a 10% lanthanum-doped $LiTaO_3$ solution.

The three solutions then underwent sonication process for 90 min using a Branson 2510 instrument. The solutions were then dropped toward the substrate's surface placed on a spin coating rotator with a rotation speed of 3000 rpm, conducted twice. The remaining solutions were then dried at 80°C for 24 h and then characterized using FTIR and PSA. The dropped substrates were then annealed using a furnace with an increasing rate of temperature at 1.7°C/min, started from room temperature until it reaches 550°C and held constantly for 12.5 h, and then cooled down to room temperature. The formed thin films were characterized using XRD and optical ocean USB 2000 [11].

4. Results and discussion

4.1. La-doped BST thin films

Before we explain the effect of lanthanum on barium strontium titanate (BST), we first describe our earlier works regarding the effect of the addition of nonrare earth-dopant BST because one can compare the results to similar treatment of substrates with lanthanum dopant [9].

Gallium oxide-doped barium strontium titanate and tantalum oxide-doped barium titanate are of immense interest, particularly as a ferroelectric solar cell (FSC). The roughness and grain size properties of the BST grown films, which were already doped by gallium oxide and tantalum oxide, were measured at various annealing temperature. We found that the roughness and grain size properties of the material can be tailored by varying the concentration of the dopant and annealing temperature. This opens the possibility to optimize FSC performance by varying the doping concentration and annealing temperature. We also performed 3D atomic force microscopy (AFM) imaging to observe the effect of dopant addition on the surface roughness, and the result is depicted in **Figure 1**.

The figure clearly indicates an increase in peak roughness when gallium oxide is added, resulting in BGST but shows smooth mixing roughness for BTST. Specifically, the average surface roughness for the BST thin films were 10.50, 0.6324, and 0.2202 nm at annealing temperature

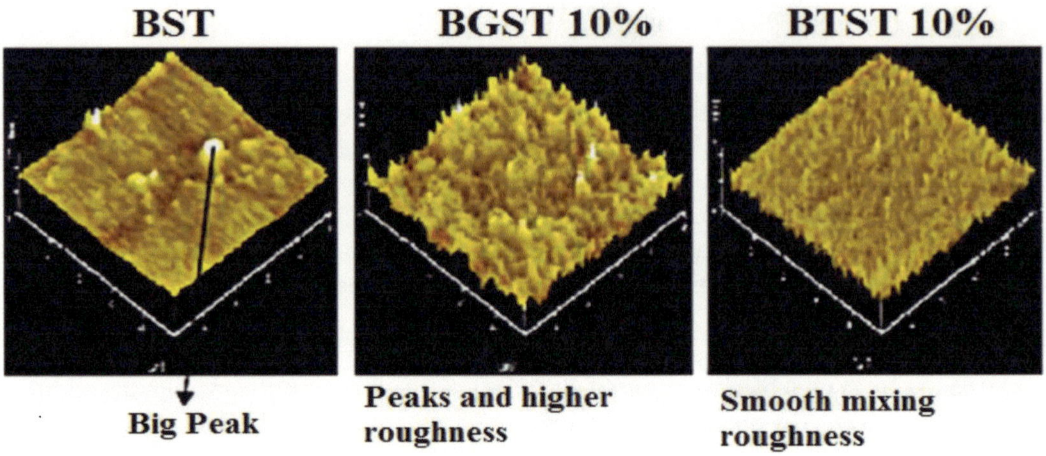

Figure 1. The three-dimensional analysis using the AFM method at 280°C of a 5000 × 5000 nm substrate area for BST, BGST, and BTST thin films [9].

of 200, 240, and 280°C, respectively, whereas the grain sizes (mean diameter) were 1.046, 238.4, and 141.3 nm, respectively. Meanwhile, the average surface roughness for BGST thin films were 0.632, 0.564, and 0.487 nm at annealing temperature of 200, 240, and 280°C, respectively, whereas the grain sizes (mean diameter) were 238.4, 219.0, and 185.1 nm, respectively. The average surface roughness for BTST thin films were 1.087, 0.4870, and 0.2317 nm at temperature 200, 240, 280°C, respectively, whereas the grain size (mean diameter) are 158.7, 291.1, and 396.7 nm, respectively [9]. Thus, we found that addition of dopants significantly alters the surface roughness and grain size, and a similar feature is expected for rare-earth, e.g., lanthanum oxide (La_2O_3) dopant on thin films.

The optical and structural effects of lanthanum oxide (La_2O_3) dopant on ferroelectric barium strontium titanate (BST) thin film were performed in our research. As expected from our previous work [9], we observe a decrease in the lattice parameter of La-doped BST thin films (BSLT) compared to nondoped thin films which were 3.936 and 3.949 Å, respectively. The reason for this decrease is due to the insertion of La atoms as substitutes in the BST lattice. We attribute the process more specifically as follows: the element lanthanum which has a radius of 1.15 Å replaces one Sr^{2+} ion in the BST structure which has radius of 1.13 Å and not the Ba^{2+} and Ti^{4+} ions which have radii of 1.35 and 0.68 Å, respectively. This preference is due to the close similarity of their atomic radii. Meanwhile, an electrostatic imbalance occurs which we attribute to an excess La^{3+} proton, which replaces one of other Sr^{2+} ions, resulting in a repulsive force with other Sr^{2+} ions in the crystal structure.

In addition, the lanthanum dopant significantly decreases the absorbance value, which is shown in **Figure 2**. This decrease in absorbance is due to the addition of La and can be attributed to the substation of La on the Sr ion which in turn changes the eigen energy level of the BSLT thin film. As shown in **Figure 3**, the insertion of La in BST increases the bandgap of the system. The result shows that the energy gap for pure BST was 1.99 and 2.67 eV for a 6% La-doped BST. The energy gap was obtained using [12]

$$\left(\frac{\alpha hc}{\lambda}\right)^2 = \text{Cons.}\left(\frac{hc}{\lambda} - E_g\right) \quad (3)$$

where α is the absorption coefficient, (hc) is the incident photon energy, and E_g is the band-gap energy. The energy gap can be obtained via linear plotting $\left(\frac{\alpha hc}{\lambda}\right)$ versus E_g. As shown in **Figure 2**, both pure BST and BSLT thin films have good absorption in the range of 400–700 nm which is in the visible light spectrum. Therefore, it is not surprising that we find a decrease in the absorbance spectrum since the electrons are now unable to absorb photons with energy less than the bandgap energy as was possible for undoped BST (smaller energy gap). Again, *ab initio* quantum mechanical calculation is necessary to investigate the full-band structure of BSLT, which is beyond the scope of this work.

Furthermore, we measured and found that the addition of lanthanum oxide will increase the refractive index and energy gap of the semiconductor thin films, which is depicted in **Table 2**. These numbers were obtained using Eq. (2). This indicates that the electrons are harder to be displaced from the valence band to the conduction band.

The crystal structure of BST films was characterized using XRD. **Figure 4** shows that diffraction peaks on each sample were in accordance with the peaks of BST on JCPDF data number 39-1395. The peaks of BST were obtained at the angle of 2θ: 22.49, 45.58, 66.98, and 76.22° corresponding to Miller planes of (1 0 0), (2 0 0), (2 2 0), and (3 1 0). The addition of lanthanum on BST films results in a decrease of the diffraction peaks but also contributes to the occurrences of new peaks at a diffraction angle of 2θ: 32.09 and 76.22°, corresponding to Miller

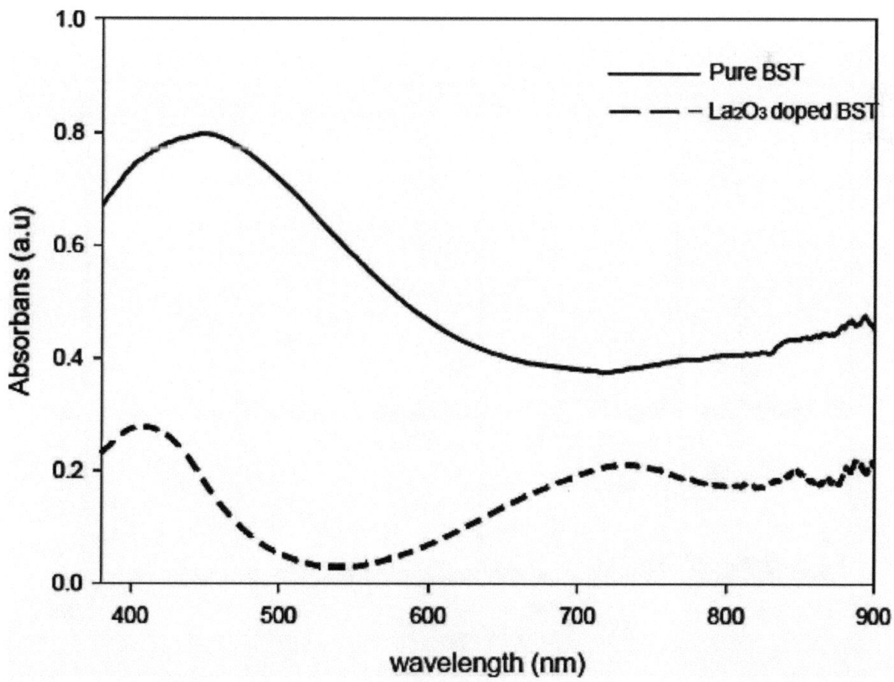

Figure 2. Absorbance curve for BST and BSLT thin films.

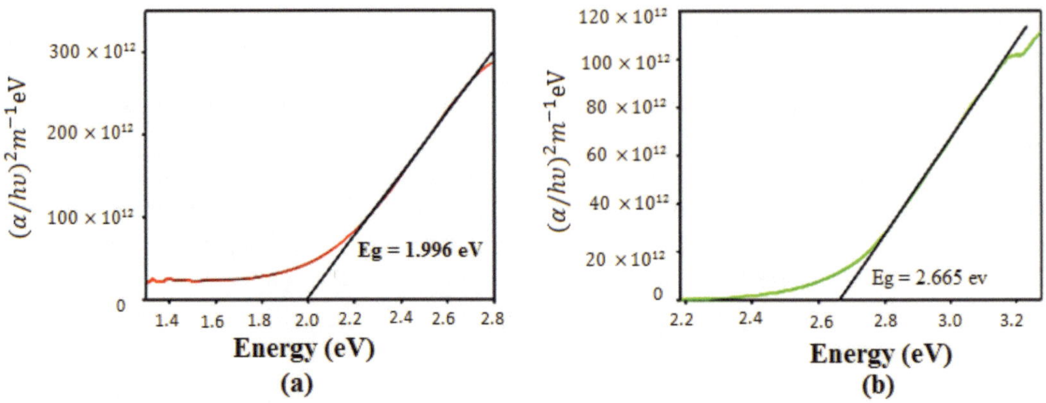

Figure 3. Energy gap of (a) undoped BST and (b) 6% La-doped BST thin films.

Film	Refractive index	Energy (eV)
BST 0%	4.386	1.996
BSLT 6%	13.118	2.665

Table 2. The refractive index and energy for BST and BSLT 6%.

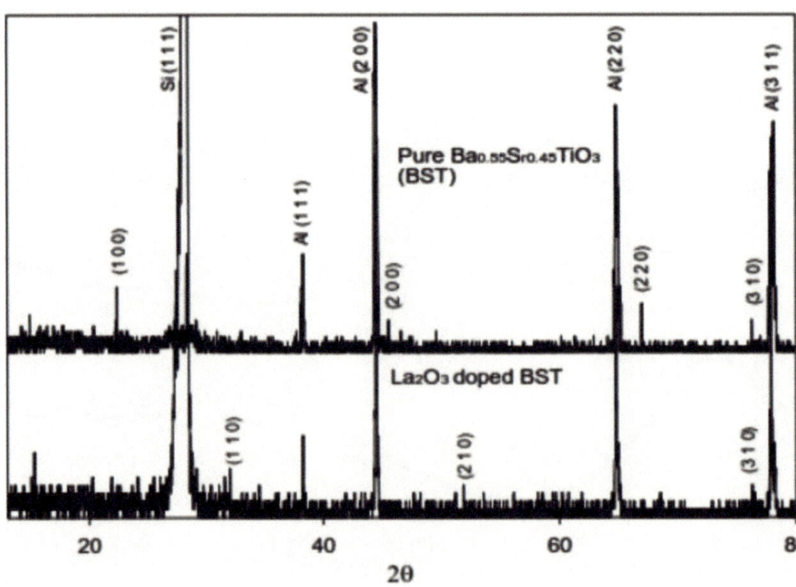

Figure 4. XRD pattern of BST and BST doped La_2O_3 (BSLT) thin films.

planes of (1 1 0) and (2 1 0). The result shows that addition of lanthanum oxide affects the lattice parameter value of BST, where the symmetry of BSLT is cubic.

Interestingly, this decrease-occurrences diffraction pattern is also obtained from the beforementioned work of Tran et al. in **Figure 1** of Ref. [6]. Specifically, their XRD pattern shows that the undoped BNKT-ST ceramic sintered at 1175°C for 2 h has a morphotropic phase boundary mixture of tetragonal and rhombohedral symmetry. However, upon addition of lanthanum, some of the peaks gradually decreased, indicating a phase transition from coexistence of rhombohedral and tetragonal phases to a pseudocubic phase. They suggest that La successfully diffused into the BNKT-ST lattice and in turn forming a homogeneous solid solution which seems to be the same as in our case.

The lattice parameters of BST and BSLT were obtained using the Rietveld method where we obtained the values $a = b = c$ and $\alpha = \beta = \gamma = 90°$ indicating that BST and BSLT films have cubic Bravais lattice. Therefore, the films can easily be grown on a diamond cubic silicon substrate due to their similar crystal structure [13].

4.2. La doped lithium niobate thin films

In addition to BST, we have also recently investigated the optical and structural properties of lanthanum-doped lithium niobate ($LiNbO_3$) thin films. Lithium niobate has a trigonal crystal system. Such a system lacks an inversion symmetry; hence, surface characterization using second harmonic nonlinear spectroscopic methods is not suggested due to radiation from within the bulk. However, some nonlinear effects do occur such as the Pockels effect can be observed. Lithium niobate possesses ferroelectrical behavior and is transparent when shined by visible light.

Figure 5 shows the relationship between reflectance and wavelength of lanthanum-doped $LiNbO_3$ thin films. It can be inferred that the dopant concentration affects the reflectance intensity, e.g., 5% lanthanum-doped films increase the reflectance intensity, but conversely, 10% lanthanum-doped thin films decrease the reflectance intensity. The reason why the reflectance intensity decreases for the 10% lanthanum doping is due to the increase in the constituent atoms, which in turn increases the collision between light particle and atoms; therefore, it becomes more difficult for the light to pass through. The less transparent the film leads to higher absorbance value and damping constant and thus the lower the transmittance value [14, 15].

The optical properties of lithium niobate have been determined by Mamoun et al. using the generalized gradient approximation (GGA) method with indirect bandgaps. They found that the energy bandgaps of $LiNbO_3$ using the GGA method were 3.32 and 3.61 eV [16]. **Figure 6** shows that the energy gap for undoped, 5, and 10% lanthanum-doped thin films were 2.75, 2.80, and 2.43 eV, respectively showing that the addition of lanthanum does not produce a consistent increase or decrease. As earlier, the energy gaps were determined using Tauc plot analysis. wRemarkably, the small increase for small La-doping concentration and decrease for higher La-doping concentration is in agreement with the results shown in **Figure 10** of Ref. [17]. The researchers argue that the observed lowering of energy gap with La doping could be due to

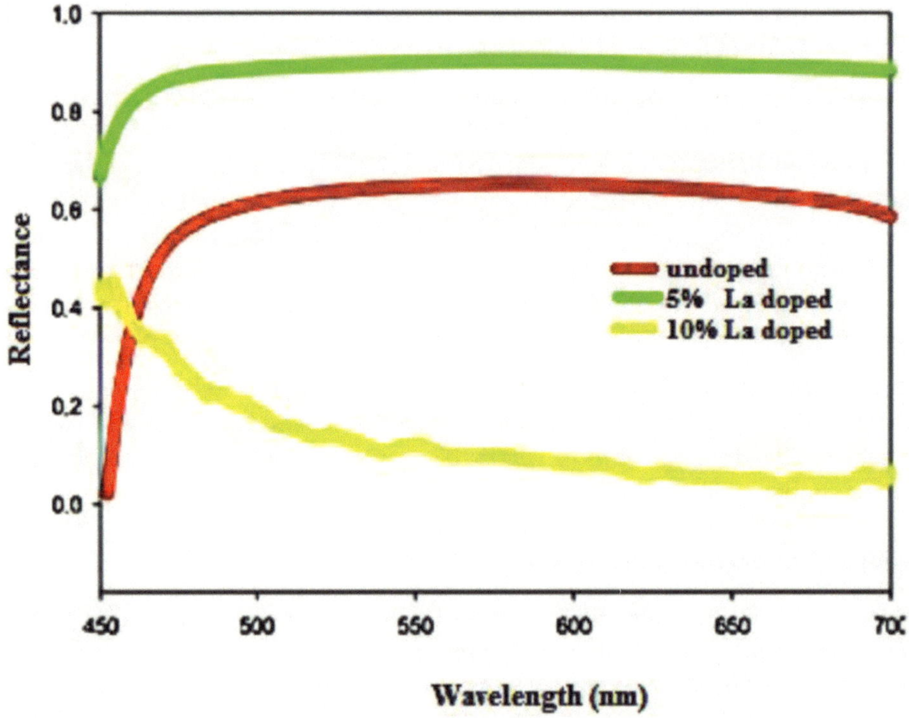

Figure 5. Reflectance of undoped and doped $LiNbO_3$ thin films at several wavelength [10].

the combined effect of increased grain size and improved crystallinity with increase in dopant concentration [17]. From the quantum mechanical point of view, the increase in the absorbance for a higher doping concentrated can be attributed to the decrease in the bandgap due to overlapping energy levels which could in principle be calculated using Density Functional Theory (DFT) or other *ab initio* methods. Up to now, the bandgap decreases—more light with less energy can be absorbed resulting in an increase in absorption. The absorption is high in the UV region and low in the IR region.

Furthermore, we also performed FTIR spectroscopic measurements on lanthanum-doped $LiNbO_3$ thin films to investigate the presence of functional groups: O-H, N-H, C=C, C-O, C-H, and hexagonal crystal symmetry. The result is present in **Figure 7**. A higher dopant concentration leads to a smaller lattice parameter. It was found that the particle sizes of undoped, 5, and 10% lanthanum doped films were 1230.59, 977.5, and 741.51 nm, respectively, thus showing a consistent decrease in the lattice parameter smaller than that of lithium niobate. The decrease in crystal size is also influenced by the radius of its constituent ions. Ionic radii of Li^+, Nb^{5+}, and La^{3+} were 0.90, 0.78, and 1.172 Å, respectively. We attribute this decrease to the following: the ionic radius of La^{3+} is closer to that of Li^+; therefore, La^{3+} can replace the positions of Li^+ in the crystal structure. Because the replaced dopant size is smaller, it leads to an overall decrease in the crystal size of doped $LiNbO_3$.

Figure 8 shows the XRD pattern of $LiNbO_3$ thin films, which are deposited on the p-type silicon (1 0 0) substrate after the annealing process at 550°C. The data were collected with interval

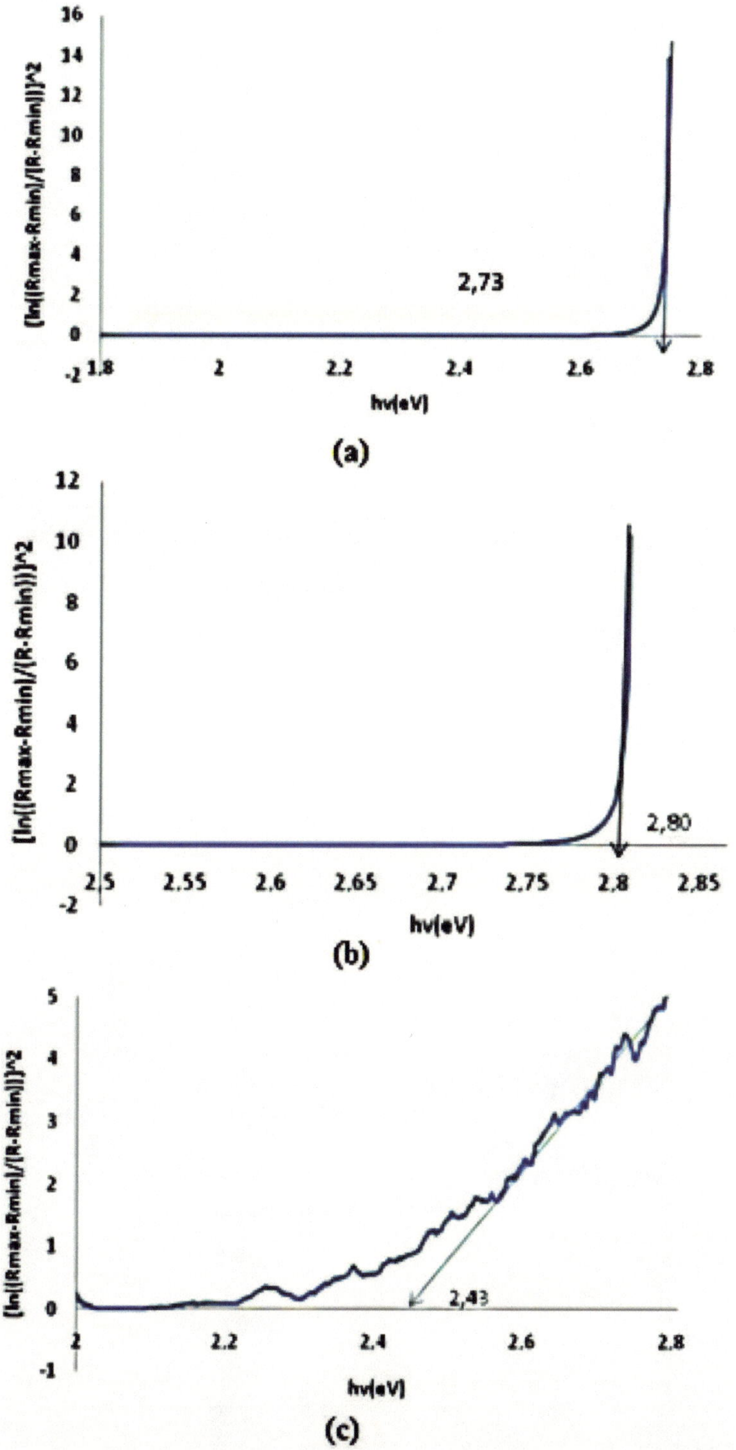

Figure 6. Tauc plot method to determine the energy gap for (a) undoped (b) 5% lanthanum doped, and (c) 10% lanthanum-doped $LiNbO_3$ thin films [10].

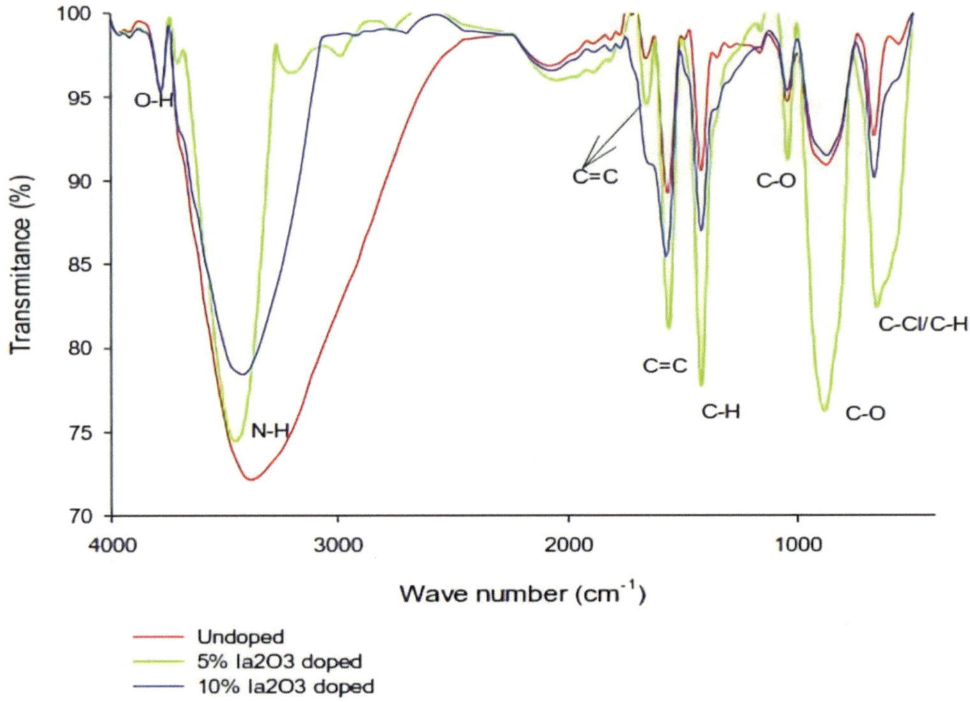

Figure 7. The FTIR spectra of $LiNbO_3$ thin films [10].

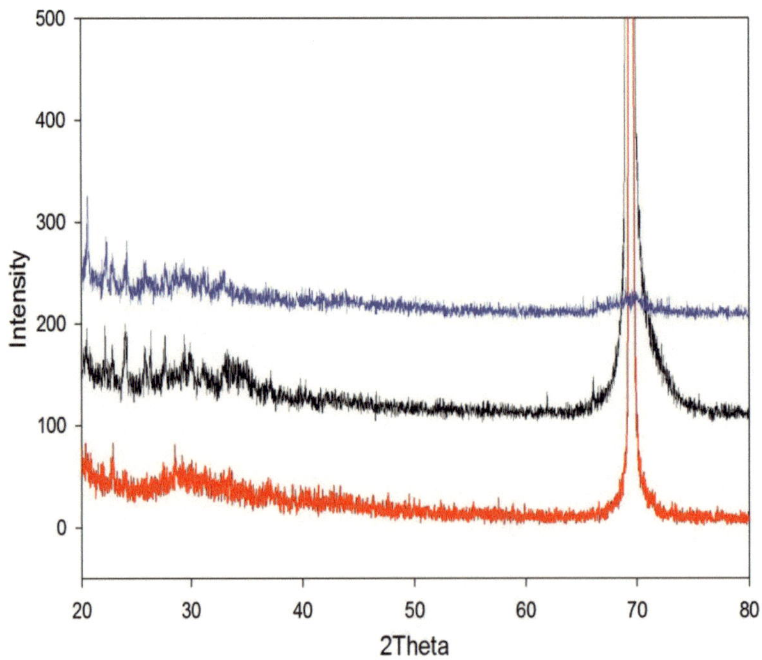

Figure 8. The XRD pattern of $LiNbO_3$ thin films [10].

http://dx.doi.org/10.5772/intechopen.69029

0.02° in the range of 10–80°C. Compared to JCPDS (Joint committee on powder diffraction standard) data, the peaks of $LiNbO_3$ are formed in the (h k l) of (0 1 2) and (1 0 4). In addition, the peaks are in the h k l of (1 0 0), (0 0 2), and (1 0 1) for thin films doped with lanthanum. The lanthanum dopant also affects the quality of $LiNbO_3$ crystal, addition of 10% dopant concentration is followed by a decrease in the crystal intensity peak at a Miller indices (hkl) of (0 1 2), while 5% dopant (blue) results in the increases the intensity peak. However, a contrary result is obtained for the 10% La dopant. The lattice parameters of $LiNbO_3$ and lanthanum were calculated using the Cohen method [13]. From the calculation and database, it is known that lattice parameter of lanthanum is smaller than that of lithium niobate, thus it can be said that the dopant addition will decrease the lattice parameter of $LiNbO_3$.

Very recently, we also performed 3D AFM imaging similar to the previous work on doped BST to investigate the surface roughness. As expected, increase in La doping results in a surface roughening as depicted in **Figure 9**. We attribute this roughening due to local strain effects due to substitution atoms, which will affect the growth kinetic process. However, an increase of dopants up to an optimal value may well lower the roughness if the spatial variations of the local strain can become less. Further work is underway to understand the fundamental physical process regarding this roughening, and the result will be reported elsewhere.

4.3. La-doped lithium tantalite thin films

In the following section, we report the effect of La doping on lithium tantalite ($LiTaO_3$) thin films, which is based on our previous work [11]. $LiTaO_3$ is one of the widely used optoelectronics materials due to its ferroelectric, piezoelectric, and pyroelectric properties. It is also an attractive material for integrated-optic applications due to its nonlinear optical properties, large electroptic and piezoelectric coefficient and its superior resistance to laser-induced optical damage. Thus, $LiTaO_3$ is a ferroelectric crystal that undergoes high Curie temperature of 608°C, and it also has high melting temperature at 1650°C [15]. **Figure 10** shows that the energy gaps of undoped, 5% La doped, and 10% La-doped thin films are 2.550, 2.020, and 2.199 eV, respectively.

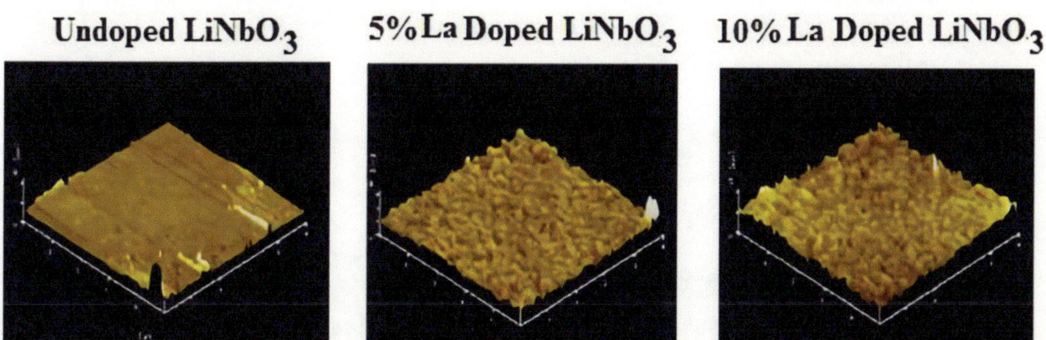

Figure 9. 3D AFM surface roughness of undoped LiNbO, 5% La doped LiNbO, and 10% LiNbO [10].

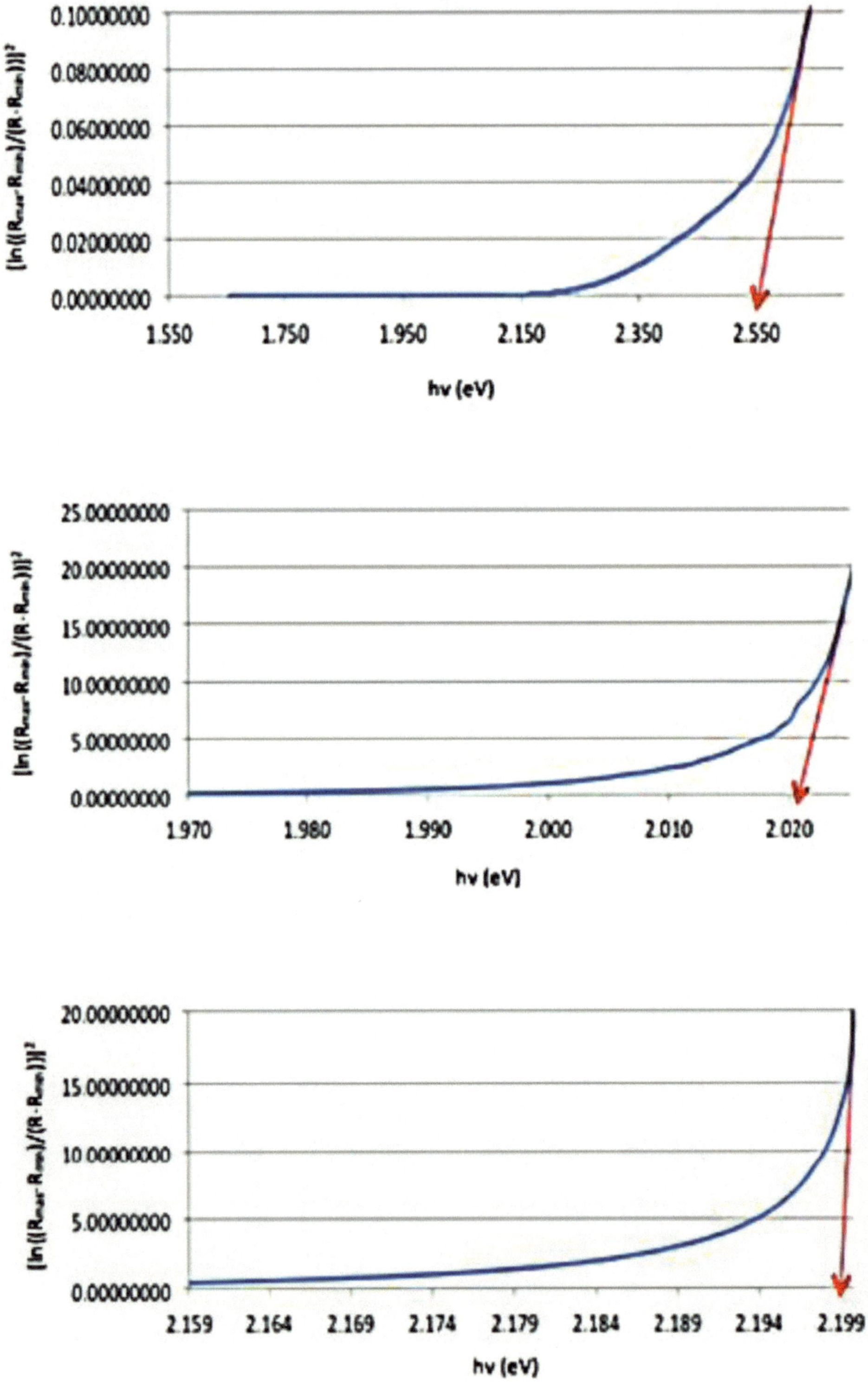

Figure 10. Energy gap of (a) undoped, (b) 5% doped, and (c) 10% doped $LiTaO_3$ thin films [11].

It can be concluded that the higher dopant concentration leads to the lower energy gap, which means the electron will be easier to move from the valence band area into the conduction band.

The energy gap in lithium tantalite is smaller than that obtained in lithium niobate. In this research, the overall values of energy gap of thin films from silicon substrate are in the range of 2.0–3.45 eV for all concentrations. *Ab initio* quantum mechanical calculations focusing on overlapping of eigen energy levels between the La upon Na and Ta energy bands should be performed to investigate the energy gap more comprehensively, but it is beyond the scope of our work.

It is well known that the energy gap of pure silicon is typically 0.7 eV. Silicon itself is usually used as a semiconductor material and has been applied on many electronic devices due to its unique energy gap. We found that there is energy gap increase for a La dopant concentration between 5 and 10%. This is in accordance with the other researcher-related work, which also found the energy gap to increase after dopant addition [18, 19]. Indeed, this result is known as the Burstein-Moss effect, which is the increase in the bandgap value of a thin film–doped semiconductor due to electrons in the conduction band shifting the Fermi level and causes an increasing bandgap [20]. The phenomenon of which the apparent bandgap of a semiconductor is increased as the absorption edge is pushed to higher energies is due to of all states that lie close to the conduction band being already populated.

Figure 11 shows the XRD pattern of $LiTaO_3$ thin films with the p-type Si (1 0 0) substrate after annealing at 550°C. The data were obtained within an interval of 0.02° in the range of 20–80°C. These three patterns show a peak at $2\theta = 70°$ corresponding to the silicon substrate. The XRD pattern shows that $LiTaO_3$ thin films have both crystalline and amorphous

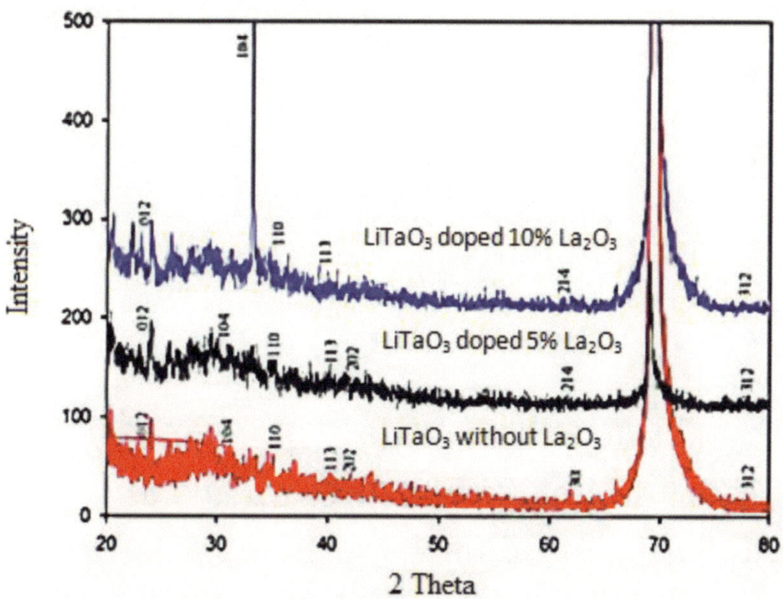

Figure 11. The XRD pattern of $LiTaO_3$ thin film with p-type Si (1 0 0) substrate after annealing at 550°C. Li^+ in the crystal structure [11].

structures. The undoped $LiTaO_3$ thin films have mixed structure with lattice parameters a = 5.13 Å and c = 13.54 Å. Meanwhile, for the 5% La-doped $LiTaO_3$ thin films, the XRD pattern shows that the structure is not significantly different compared to that of without doping albeit smaller lattice parameters: a = 4.92 Å and c = 14.19 Å. Moreover, for the 10% La_2O_3 doped $LiTaO_3$ thin films, a peak of $LiTaO_3$ with a lattice parameter of a = 5.11 Å and c = 13.30 Å is present. We also argue that the decrease in crystal size is influenced by the radii of its constituent ions.

We measure the ionic radii of Li^+, Ta^{5+}, and La^{3+} to be 0.90, 0.78, and 1.172 Å, respectively. The ionic radius of La^{3+} is closer to that of Li+; therefore, we propose that the La^{3+} ion can occupy the positions of the La-doped $LiTaO_3$ thin films. The difference of ionic radii between dopant and replaced ion affects the formation of spinel phase. This leads to crystal size decreasing due to the existence of dopant cations in the $LiTaO_3$ structure. Furthermore, investigation of the composition using FTIR shows the existence of stretching vibration of OH group, C=C aromatic bonding, Li–O bonding, and Ta–O bonding at wave numbers of 3100–3900, 1650–1450, 1440–1420, and 610–945 cm^{-1}, respectively. The FTIR result is depicted in **Figure 12**.

The addition of 5% La-dopant significantly increases the refractive index compared to the undoped thin films. However, at a 10% La dopant, the refractive index decreases with respect to the undoped, as shown in **Figure 13**, the physical reason of this decrease is still not known to us. Further work needs to be performed to explain this result, e.g., studying the band structure effect due to dopant at various wavelengths.

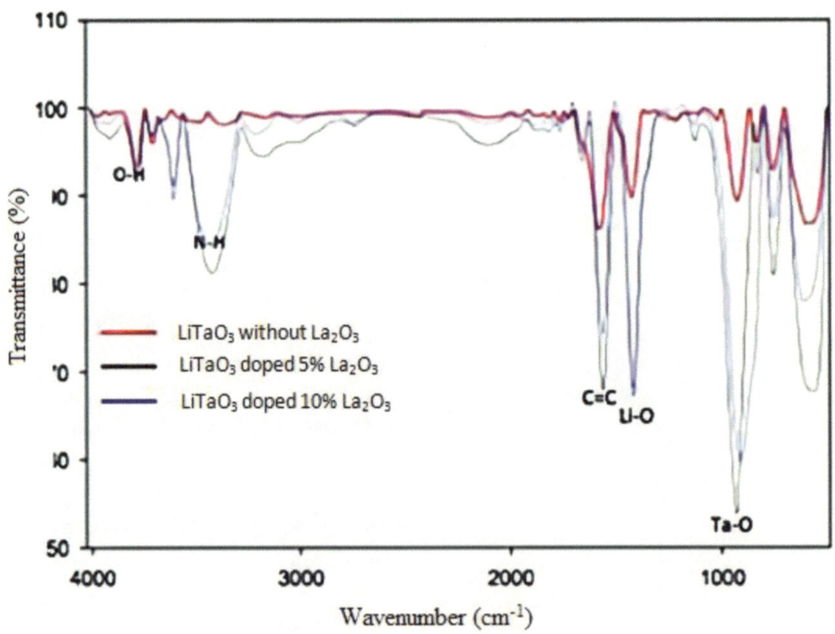

Figure 12. FTIR spectra of La doped and undoped $LiTaO_3$ [11].

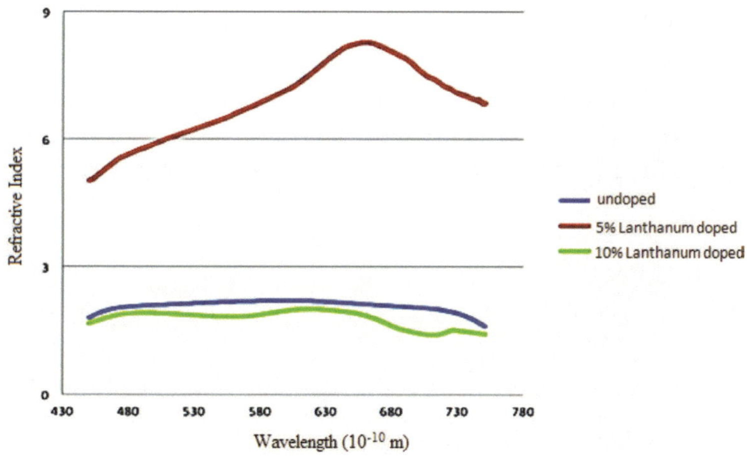

Figure 13. Refractive index variation due to La doping on $LiTaO_3$ [11].

5. Conclusion

We have showed that the effect of lanthanum doping on a ferroelectric thin film produces significant changes in its structural and optical property. The optical and structural effects of lanthanum oxide (La_2O_3) dopant on ferroelectric barium strontium titanate (BST) thin film results in a decrease in the lattice parameter of La-doped BST thin films (BSLT) compared to nondoped thin films. The addition of lanthanum oxide dopant on BST films results in a decrease of the diffraction peaks but also contributes to the occurrences of new peaks at a diffraction angle. The reason for this decrease is due to the insertion of La atoms as substitutes of Sr^{2+} in the BST lattice. This preference is due to the close similarity of their atomic radii. Meanwhile, addition of lanthanum dopant on lithium niobate thin films affects the reflectance intensity, e.g., 5% lanthanum doped increase the reflectance intensity, but conversely, 10% lanthanum-doped thin films decrease the reflectance intensity. From the quantum mechanical point of view, the increase in the absorbance for a higher doping concentrated can be attributed to the decrease in the bandgap due to overlapping energy levels, which could in principle be calculated using DFT or other *ab initio* methods. The 3D AFM imaging on La-doped lithium niobate produced similar results as for doped BST, namely, an increase in the surface roughening due to dopant. Finally, the addition of lanthanum on lithium tantalite shows the energy gap decrease of La dopant thin films compared to the undoped but increase slightly for the 10% La concentration. This is in accordance with other researcher-related work, which also found the energy gap to increase after continuous dopant addition confirming the Burstein-Moss effect. The phenomenon of which the apparent bandgap of a semiconductor is increased as the absorption edge is pushed to higher energies is due to of all states that lie close to the conduction band being already populated thus shifting up the Fermi level.

Acknowledgements

This work was funded by the International Research Collaboration and Scientific Publication under contract No. 082/SP2H/UPL/DIT.LITABMAS/II/2015 and Penelitian Institusional Institut Pertanian Bogor in 2016, Ministry of Research, Technology, and Higher Education, Republic of Indonesia.

Author details

Irzaman[1*], Hendradi Hardhienata[1], Akhiruddin Maddu[1], Aminullah[2] and Husin Alatas[1]

*Address all correspondence to: irzaman@apps.ipb.ac.id

1 Department of Physics, Faculty of Mathematics and Natural Sciences, Bogor Agricultural University of Indonesia, Bogor, Indonesia

2 Department of Food Technology and Nutrition, Djuanda University of Indonesia, Bogor, Indonesia

References

[1] Smith RJ, Piacentini M, Lynch DW. Multiplet structure below "Threshold" in appearance-potential spectra—lanthanum $N_{4,5}$. Physical Review Letters. 1975;**34**(8):476-479

[2] Kanski J, Nilsson PO, Curelaru I. Studies of the 4f states in lanthanum by means of electron and photon excited appearance potential spectroscopy. Journal of Physics F: Metal Physics. 1976;**6**:1073-1077

[3] Moore RJ. Lanthanum: Compounds, Production and Applications, Chemistry Research and Applications. New York: Nova Science Publishers; 2011. pp. 349-363

[4] Cai W, Fu C, Gao J, Zhao C. Dielectric properties and microstructure of Mg doped barium titanate ceramics. Advances in Applied Ceramics. 2011;**110**(3):181-185

[5] Falcao EA, Eiras JA, Garcia D, Santos IA, Medina AN, Baesso ML, Catunda T, Guo R, Bhalla AS. Effects of lanthanum content on the thermo-optical properties of (Pb,La)(Zr,Ti)O_3. Ferroelectrics. 2016;**494**(1):33-42

[6] Tran VDN, Ullah A, Dinh TH, et al. Effect of lanthanum doping on ferroelectric and strain properties of $0.96Bi_{1/2}(Na_{0.84}K_{0.16})_{1/2}TiO_3$-$0.04SrTiO_3$ lead-free ceramics. Journal of Electronic Materials. 2016;**45**:2639-2643

[7] Han JY, Bark CW. Tunable band gap of iron-doped lanthanum-modified bismuth titanate synthesized by using the thermal decomposition of a secondary phase. Journal of the Korean Physical Society. 2015;**66**:1371-1375

[8] Bae J, Kimb SS, Choib EK, Songc TK, Kimb W-J, Lee Y-I. Ferroelectric properties of lanthanum-doped bismuth titanate thin films grown by a sol–gel method. Thin Solid Films. 2005;**472**(1-2):90-94

[9] Irzaman, Darmasetiawan H, Hardhienata H, Hikam M, Arifin P, Jusoh, S. Taking SN, Jamal Z, Idris MA. Electrical properties of photodiode $Ba_{0.25}Sr_{0.75}TiO_3$ (BST) thin film doped with ferric oxide on p-type Si (1 0 0) substrate using chemical solution deposition method. Atom Indonesia. 2009;**35**(1):133-138

[10] Irzaman, Sitompul H, Masitoh, Misbakhusshudur M, Mursyidah. Optical and structural properties of lanthanum doped lithium niobate thin films. Ferroelectrics. 2016;**502**(1):9-18

[11] Irzaman, Pebriyanto Y, Apipah ER, Noor I, Alkadri A. Characterization of optical and structural of lanthanum doped $LiTaO_3$ thin films. Integrated Ferroelectrics. 2015;**167**(1): 137-145

[12] Itskovsky MA. Kinetics of ferroelectric phase transition: Nonlinear pyroelectric effect and ferroelectric solar cell. Japanese Journal of Applied Physics. 1999;**38**(8):4812-4817

[13] Wang SY, Cheng BL, Wang C, Dai SY, Lu HB, Zhou YL, Chen ZH, Yang GZ. Raman spectroscopy studies of Ce-doping effects on $Ba_{0.5}Sr_{0.5}TiO_3$ thin films. Applied Physics Letters. 2004;**84**:4116

[14] Shandilya S, Sreenivasa K, Katiyarb RS, Gupta V. Structural and optical studies on texture $LiNbO_3$ thin film on (0 0 0 1) sapphire. Indian Journal of Engineering & Materials Science. 2008;**15**:355-357

[15] Irzaman, Syafutra H, Arif A, Alatas H, Hilaluddin MN, Kurniawan A, Iskandar J, Dahrul M, Ismangil A, Yosman D, Aminullah, Prasetyo LB, Yusuf A, Kadri TM. Formation of solar cells based on Ba0.5 Sr0.5 TiO_3 (BST) ferroelectric thick film. 5th Nanoscience and Nanotechnology Symposium (NNS2013). AIP Conference Proceedings. 2014;**1586**:24-34.

[16] Mamoun S, Merad AS, Guilbert L. Electronic and optical properties of lithium niobate from density functional theory. Computational Materials Science. 2013;**79**:125-131

[17] Majumder SB, Jain M, Katiyar RS. Investigations on the optical properties of sol–gel derived lanthanum doped lead titanate thin films. Thin Solid Films. 2002;**402**:90-98

[18] Ahmed AS, Muhamed Shafeeq M, Singla ML, Tabassum S, Naqvi AH, Azam A. Band gap narrowing and fluorescence properties of nickel doped SnO_2 nanoparticles. Journal of Luminescence. 2011;**131**:1-6

[19] Chakraborty M, Ghosh A, Thangavel R. Experimental and theoretical investigations of structural and optical properties of copper doped ZnO nanorods. Journal of Sol-Gel Science and Technology. 2015;**74**:756-764

[20] Manser JS, Kamat P. Band filling with free charge carriers in organometal halide perovskites. Nature Photonics. 2014;**8**(9):737-743